Ecology and Applied Environmental Science

Ecology and Applied Environmental Science

Contributors

Mohammad Ali Zare Chahouki et al.

www.aurisreference.com

Ecology and Applied Environmental Science

Contributors: Mohammad Ali Zare Chahouki et al.

Published by Auris Reference Limited
www.aurisreference.com

United Kingdom

Copyright 2016
Printed in 2017 for Sale in the Indian Subcontinent

The information in this book has been obtained from highly regarded resources. The copyrights for individual articles remain with the authors, as indicated. All chapters are distributed under the terms of the Creative Commons Attribution License, which permit unrestricted use, distribution, and reproduction in any medium, provided the original author and source are credited.

Notice

Contributors, whose names have been given on the book cover, are not associated with the Publisher. The editors and the Publisher have attempted to trace the copyright holders of all material reproduced in this publication and apologise to copyright holders if permission has not been obtained. If any copyright holder has not been acknowledged, please write to us so we may rectify.

Reasonable efforts have been made to publish reliable data. The views articulated in the chapters are those of the individual contributors, and not necessarily those of the editors or the Publisher. Editors and/or the Publisher are not responsible for the accuracy of the information in the published chapters or consequences from their use. The Publisher accepts no responsibility for any damage or grievance to individual(s) or property arising out of the use of any material(s), instruction(s), methods or thoughts in the book.

Ecology and Applied Environmental Science

ISBN: 978-1-78154-981-0

British Library Cataloguing in Publication Data
A CIP record for this book is available from the British Library

Printed in the United Kingdom

Exclusively distributed by CBS Publishers & Distributors Pvt. Ltd.

Sales & Distribution Rights only for India, Pakistan, Bangladesh, Sri Lanka, Nepal and Bhutan. This book is not to be sold outside these territories.

Contents

List Of Abbreviations vii
List Of Contributors... ix
Preface.. xv

Chapter 1　Multivariate Analysis Techniques in Environmental Science 1
Mohammad Ali Zare Chahouki

Chapter 2　Contamination of Environment with Polycyclic Aromatic Hydrocarbons in India 35
Khageshwar Singh Patel, Shobhana Ramteke, Yogita Naik, Bharat Lal Sahu, Saroj Sharma, Jutta Lintelmann, Matuschek Georg

Chapter 3　Environment and Health in Contaminated Sites: The Case of Taranto, Italy 51
Roberta Pirastu, Pietro Comba, Ivano Iavarone, Amerigo Zona, Susanna Conti, Giada Minelli, Valerio Manno, Antonia Mincuzzi, Sante Minerba, Francesco Forastiere, Francesca Mataloni, and Annibale Biggeri

Chapter 4　The Mexican Environmental Flow Standard: Scope, Application and Implementation 87
María Antonieta Gómez-Balandra, María del Pilar Saldaña-Fabela, Maricela Martínez-Jiménez

Chapter 5　Evaluation of Soil Hydraulic Parameters in Soils and Land Use Change.. 103
Fereshte Haghighi, Mirmasoud Kheirkhah and Bahram Saghafian

Chapter 6　Energy-Related Carbon Emissions of China's Model Environmental Cities 115
Kevin Lo

Chapter 7　Cultivating a Value for Non-Human Interests through the Convergence of Animal Welfare, Animal Rights, and Deep Ecology in Environmental Education 129
Helen Kopnina, and Brett Cherniak

Chapter 8	**An Ecology for Cities: A Transformational Nexus of Design and Ecology to Advance Climate Change Resilience and Urban Sustainability** 151	

Daniel L. Childers, Mary L. Cadenasso, J. Morgan Grove, Victoria Marshall, Brian McGrath and Steward T. A. Pickett

Chapter 9 **Industrial Ecology and Environmental Lean Management: Lights and Shadows** 175

Giuseppe Ioppolo, Stefano Cucurachi, Roberta Salomone, Giuseppe Saija and Luigi Ciraolo

Chapter 10 **Ecology and eScience** 195

Christian Mulder

Chapter 11 **A Bibliometric Analysis of Global Forest Ecology Research during 2002–2011** 205

Yajun Song and Tianzhong Zhao

Chapter 12 **Size-Energy Relationships In Ecological Communities** 225

Brent J. Sewall, Amy L. Freestone, Joseph E. Hawes, Ernest Andriamanarina

Chapter 13 **Seismic Microzonation Of Breginjski Kot (Nw Slovenia) Based On Detailed Engineering Geological Mapping** 253

Jure Kokošin and Andrej Gosar

Citations 275

Index 279

List of Abbreviations

AFOLU	Agriculture, forest management, and other land uses
ALF	Animal Liberation Front
AR	Animal rights
AW	Animal welfare
AWE	Animal welfare education
BES LTER	Baltimore Ecosystem Study Long-Term Ecological Research Program
BHC	Building Healthy Communities
BD	Bulk density
CCA	Canonical correspondence analysis
CCA	Canonical Correspondence Analysis
CC	Carbonate carbon
CAM	Computer Aided Manufacturing
CIM	Computer Integrated Manufacturing
CAFOs	Concentrated animal feeding operations
CRW	correlated random walk
DE	Deep ecology
DI	Deprivation index
ELF	Earth Liberation Front
ENA	Ecological network analysis
ESD	Education for sustainable development
EC	Elemental carbon
ERP	Enterprise Resource Planning
EE	Environmental education
EFA	Environmental flow assessment
EIA	Environmental Impact Assessment
EO	Environmental objective
ETP	European Technology Platforms
FC	Field capacity
FDE	Foundation for Deep Ecology
IE	Industrial Ecology
ICT	Information Communication Technology
IWRM	Integrated Water Resources Management
IEA	International Energy Agency
IFAW	International Fund for Animal Welfare
JIT	Just-in-time
KLT	Karhunen–Loève transform
LM	Lean Management
LP	Lean Production
MAF	Mean annual flow
MRPP	Multi-response Permutation Procedure
NDRC	National Development and Reform Commission
NPCSs	National Priority Contaminated Sites
NSF	National Science Foundation
NMS	Non-metric Multidimensional Scaling
OC	Organic carbon
PTFs	Pedo transfer functions

PETA	People for the Ethical Treatment of Animals
PWP	Permanent wilting point
PIOT	Physical input-output table
PAHs	Polycyclic aromatic hydrocarbons
PEU	population energy use
PCA	Principal component analysis
PCoA	Principal Coordinate Analysis
POD	Proper orthogonal decomposition
RDA	Redundancy analysis
RSPCA	Royal Society for the Prevention of Cruelty to Animals
SCEU	size class energy use
SEP	Socioeconomic position level
SOM	Soil organic matter
SDRs	Standardized death rates
SIR	Standardized incidence ratio
SMR	Standardized mortality ratios
TC	Total carbon
TOVR	Total Ordinary Volume Regime
TQM	Total quality management
TVFR	Total Volume for Flood Regime
TEF	Toxic equivalent factor
TPS	Toyota Production System
UDWG	Urban Design Working Group

List of Contributors

Mohammad Ali Zare Chahouki
Department of Rehabilitation of Arid and Mountainous Regions, University of Tehran, Iran

Khageshwar Singh Patel
School of Studies in Chemistry/Environmental Science, Pt. Ravishankar Shukla University, Raipur, India

Shobhana Ramteke
School of Studies in Chemistry/Environmental Science, Pt. Ravishankar Shukla University, Raipur, India

Yogita Naik[1],
School of Studies in Chemistry/Environmental Science, Pt. Ravishankar Shukla University, Raipur, India

Bharat Lal Sahu
School of Studies in Chemistry/Environmental Science, Pt. Ravishankar Shukla University, Raipur, India

Saroj Sharma
Department of Chemistry, Devi Rathi Mahila Mahavidhaya, Rajnandagaon, India

Jutta Lintelmann
GSF-Forschungszentrum für Umwelt und Gesundheit, Institut für Ökologische Chemie, Neuherberg, Germany

Matuschek Georg
GSF-Forschungszentrum für Umwelt und Gesundheit, Institut für Ökologische Chemie, Neuherberg, Germany

Roberta Pirastu
Department of Biology and Biotechnologies Charles Darwin, Sapienza Rome University, Piazzale Aldo Moro 5, 00185 Rome, Italy

Pietro Comba
Department of Environment and Primary Prevention, National Health Insti-

tute, Viale Regina Elena 299, 00161 Rome, Italy

Ivano Iavarone
Department of Environment and Primary Prevention, National Health Institute, Viale Regina Elena 299, 00161 Rome, Italy

Amerigo Zona
Department of Environment and Primary Prevention, National Health Institute, Viale Regina Elena 299, 00161 Rome, Italy

Susanna Conti
Unit of Statistics of the National Health Institute, National Center for Epidemiology, Surveillance and Health Promotion, Viale Regina Elena 299, 00161 Rome, Italy

Giada Minelli
Unit of Statistics of the National Health Institute, National Center for Epidemiology, Surveillance and Health Promotion, Viale Regina Elena 299, 00161 Rome, Italy

Valerio Manno
Unit of Statistics of the National Health Institute, National Center for Epidemiology, Surveillance and Health Promotion, Viale Regina Elena 299, 00161 Rome, Italy

Antonia Mincuzzi
Taranto Local Health Unit, Epidemiological and Statistical Unit, Viale Virgilio 31, 74121 Taranto, Italy

Sante Minerba
Taranto Local Health Unit, Epidemiological and Statistical Unit, Viale Virgilio 31, 74121 Taranto, Italy

Francesco Forastiere
Department of Epidemiology, Lazio Regional Health Service, Via di Santa Costanza 53, 00198 Rome, Italy

Francesca Mataloni
Department of Epidemiology, Lazio Regional Health Service, Via di Santa Costanza 53, 00198 Rome, Italy

Annibale Biggeri
Biostatistics Unit, ISPO Cancer Research and Prevention Institute, Via Cosimo il Vecchio 2, 50139 Florence, Italy
Annibale Biggeri Department of Statistics "G. Parenti", University of Florence, Viale Morgagni 59, 50134 Firenze, Italy

María Antonieta Gómez-Balandra,
Hydrobiology and Environmental Assessment Department, Mexican Institute of Water Technology, Jiutepec, México

María del Pilar Saldaña-Fabela,
Hydrobiology and Environmental Assessment Department, Mexican Institute of Water Technology, Jiutepec, México

Maricela Martínez-Jiménez
Hydrobiology and Environmental Assessment Department, Mexican Institute of Water Technology, Jiutepec, México

Fereshte Haghighi
Soil Conservation and Watershed Management Institute, Tehran

Mirmasoud Kheirkhah
Soil Conservation and Watershed Management Research Institute, Tehran, Iran

Bahram Saghafian
Soil Conservation and Watershed Management Research Institute, Tehran, Iran

Kevin Lo
Department of Resource Management and Geography, University of Melbourne, Melbourne, VIC 3010, Australia

Helen Kopnina
Institute Cultural Anthropology and Development Sociology, Social and Behavioural Sciences, Leiden University, Wassenaarseweg 52, Leiden 2300 RB, The Netherlands

Brett Cherniak
47 Jones Street, Hamilton, Ontario, ON L8R 1X9, Canada

Daniel L. Childers
School of Sustainability, Arizona State University, Tempe, AZ 85287, USA

Mary L. Cadenasso
Department of Plant Sciences, University of California, Davis, CA 95616, USA

J. Morgan Grove
U.S. Forest Service, Baltimore, MD 21201, USA

Victoria Marshall
Parsons The New School for Design, New York, NY 10011, USA

Brian McGrath
Parsons The New School for Design, New York, NY 10011, USA

Steward T. A. Pickett
Cary Institute of Ecosystem Studies, Millbrook, NY 12545, USA

Giuseppe Ioppolo
Department of Economics, Business, Environment and Quantitative Methods (SEAM), University of Study of Messina, P.zza Puglatti 1, Messina 98122, Italy

Stefano Cucurachi
Department CML-Industrial Ecology, Leiden University, 2300 RA Leiden, The Netherlands

Roberta Salomone
Department of Economics, Business, Environment and Quantitative Methods (SEAM), University of Study of Messina, P.zza Puglatti 1, Messina 98122, Italy

Giuseppe Saija
Department of Economics, Business, Environment and Quantitative Methods (SEAM), University of Study of Messina, P.zza Puglatti 1, Messina 98122, Italy

Luigi Ciraolos
Department of Economics, Business, Environment and Quantitative Methods (SEAM), University of Study of Messina, P.zza Puglatti 1, Messina 98122, Italy

Christian Mulder
National Institute for Public Health and the Environment (RIVM)

Yajun Song
School of Information Science & Technology, Beijing Forestry University
Library of Beijing International Studies University

Tianzhong Zhao
School of Information Science & Technology, Beijing Forestry University

Brent J. Sewall
Department of Biology, Temple University, Philadelphia, Pennsylvania, United States of America
Department of Wildlife, Fish, and Conservation Biology, University of California Davis, Davis, California, United States of America

Amy L. Freestone
Department of Biology, Temple University, Philadelphia, Pennsylvania, United States of America

Joseph E. Hawes
School of Environmental Sciences, University of East Anglia, Norwich Research Park, Norwich, United Kingdom

Ernest Andriamanarina
De´partement des Sciences de la Nature et de l'Environnement, Universite´ d'Antsiranana, Antsiranana, Madagascar

Jure Kokošin
Sweco Norge AS, Dronningensgate 52/54, 8509 Narvik, Norway

Andrej Gosar
Faculty of Natural Sciences and Engineering, University of Ljubljana, Aškerčeva 12, SI-1000 Ljubljana, Slovenia
Slovenian Environment Agency, Seismology and Geology Office, Dunajska 47, SI-1000 Ljubljana, Slovenia

Ichiro Ken Shimatani
The Institute of Statistical Mathematics, Tokyo, Japan

Ken Yoda
Graduate School of Environmental Studies, Nagoya University, Furo, Chikusa, Nagoya, Japan

Nobuhiro Katsumata
International Coastal Research Center, Atmosphere and Ocean Research Institute, The University of Tokyo, Kashiwa, Chiba, Japan

Katsufumi Sato
International Coastal Research Center, Atmosphere and Ocean Research Institute, The University of Tokyo, Kashiwa, Chiba, Japan

Preface

Ecology is the scientific analysis and study of interactions among organisms and their environment. The text Ecology and Applied Environmental Science addresses the impact of contemporary environmental problems by using the main principles of scientific ecology. It presents a variety of scientific ecological issues and examine a range of environmental problems while considering potential engineering, scientific, and managerial solutions. Multivariate analysis techniques in environmental science have been presented in first chapter. In second chapter, we discuss the contamination of environment with polycyclic aromatic hydrocarbons in India. Third chapter presents an epidemiological profile of Taranto NPCS residents analyzing different health indicators available at municipality level. In fourth chapter, an analysis of the environmental flow standard and examples of the suggested hydrological methods are presented. In fifth chapter, we review soil water content, pedotransfer functions, and some infiltration models applicability for two land-use types. Sixth chapter identifies three types of model environmental cities in China and examines their levels of energy-related carbon emissions using a bottom-up accounting system. Seventh chapter discusses how to cultivate a value for non-human interests through the convergence of animal welfare, animal rights, and deep ecology in environmental education. In eighth chapter, we present a transformative model that merges urban design and ecology into an inclusive, creative, knowledge-to-action process. The aim of ninth chapter is to highlight the key factors relevant to their integration in an environmental lean management system, both of positive terms (lights) and possible barrier (shadows). Tenth chapter focuses on ecology and eScience. A bibliometric analysis of global forest ecology research during 2002–2011 has been proposed in eleventh chapter. Size-energy relationships in ecological communities have been established in twelfth chapter. Last chapter represents a contribution to the evaluation of the seismic hazard in Breginjski kot, which is among the most endangered seismic areas in Slovenia.

Chapter 1

MULTIVARIATE ANALYSIS TECHNIQUES IN ENVIRONMENTAL SCIENCE

Mohammad Ali Zare Chahouki

Department of Rehabilitation of Arid and Mountainous Regions, University of Tehran, Iran

INTRODUCTION

One of the characteristics of environmental data, many of them and the complex relationships between them. To reduce the number variables, different statistical methods exist. Multivariate statistics is used extensively in environmental science. It helps ecologists discover structure and previous relatively objective summary of the primary features of the data for easier comprehension. However, it is complicated in theoretical structure and in operational methodology.

In this chapter some important statistical methods such as Principal component analysis (PCA), Canonical correspondence analysis (CCA), Redundancy analysis (RDA), Cluster analysis, and Discriminate function analysis will be explained briefly. This chapter too cover the statistical analysis of assemblage data (species by samples matrices of abundance, area cover etc) and/or multi variable environmental data which arise in a wide range of applications in ecology and environmental science, from basic ecological studies (e.g. of dietary composition or population size-structure), through community-based field studies, environmental impact assessments and monitoring of largescale biodiversity change, to purely physical or chemical analyses.

The use of multivariate analysis has been extended much more widely over the past 20 years. Much more is included on techniques such as Canonical Correspondence Analysis (CCA) and Non-metric Multidimensional Scaling (NMS) and another technique to include organisms and organism-environment relationships other than vegetation. Spatially constrained data analysis will be introduced and the importance of accounting for spatial autocorrelation will be emphasized. Use of the methods within ecology and in environmental reconstruction will also be covered. A study and review of the application of multivariate analysis in biogeography and ecology is provided in: Kent, M. (2006).

LANDSCAPE ECOLOGY

Landscape is simply an area of land (at any scale) containing an interesting pattern that affects and is affected by an ecological process of interest. Landscape ecology, then, involves the study of these landscape patterns, the interactions among the elements of this pattern, and how these patterns and interactions change over time. In addition, landscape ecology involves the application of these principles in the formulation and solving of real-world problems (Turner et al, 2001).

Landscape ecology is perhaps best distinguished by its focus on: 1) spatial heterogeneity, 2) broader spatial extents than those traditionally studied in ecology, and 3) the role of humans in creating and affecting landscape patterns and process (Turner et al, 2001).

In effect the role of ecology, and especially that of vegetation science, has been mainly restricted to the evaluation of the landscape with respect to particular demands: either evaluation as an assessment of the qualities of the ecosystem or evaluation as a socioeconomic procedure intended to estimate the functions the natural environment fulfills for human societ (Van der Ploeg and Vlijm, 1978). Landscape ecology theory stresses the role of human impacts on landscape structures and functions. It also proposes ways for restoring degraded landscapes. Landscape ecology explicitly includes humans as entities that cause functional changes on the landscape (Mielke& Berry, 2001).

Landscape ecology theory includes the landscape stability principle, which emphasizes the importance of landscape structural heterogeneity in developing resistance to disturbances, recovery from disturbances, and promoting total system stability (Mielke& Berry, 2001). This principle is a major contribution to general ecological theories which highlight the importance of relationships among the various components of the landscape. Integrity of landscape components helps maintain resistance to external threats, including development and land transformation by human activity. Analysis of land use change has included a strongly geographical approach which has led to the acceptance of the idea of multifunctional properties of landscapes (Mielke et al, 1976). There are still calls for a more unified theory of landscape ecology due to differences in professional opinion among ecologists and its interdisciplinary approach (Bastian 2001).

Landscape ecology is distinguished by its focus on broader spatial extents than those traditionally studied in ecology. This stems from the anthropocentric origins of the discipline. Despite early attention to the effects of sample area on measurements, such as species-area relationships, the importance of scale was not widely recognized until the 1980's. Recognition that pattern-process relationships vary with scale demanded that ecologist give explicit consideration to scale in designing experiments and interpreting results.

It became evident that different problems require different scales of study, and that most problems require multiple scales of study. The theory of scale and hierarchy emerged as a framework for dealing with scale. The emergence of scale and hierarchy theory provided a partial theoretical framework for understanding pattern-process relationships, which became the basis for the emergence of landscape ecology as a discipline (Turner MG. 1989).

SAMPLING METHODS

Sampling design varies considerably with habitat type and specific taxonomic groups. Sampling design begins with a clear statement of the question(s) being asked. This may be the most difficult part of the procedure because the quality of the results is dependent on the nature of the original design.

If the sampling is for densities of organisms then at least five replicate samples per sampling site are needed because many statistical tests require that minimal number. Better yet, consider 20 replicates per sampling site and in some cases 50 or more. If sample replicates are less than five then bootstrapping techniques can be used to analyze the data (name). Some type of random sampling should be attempted (e.g., stratified random sampling) or a line intercept method used to estimate densities (e.g., Strong Method).

Measurements of important physical–chemical variables should be made (e.g.,temperature, salinity, sediment grain size, etc...). Field experiments need to be carried out with carefully designed controls. The correct spatial scale needs to be considered when planning experiments (Stiling, 2002). If vegetation is correlated with geomorphologic landforms, for this reason stratified random sampling method is better and this method employ (Greig-Smith, 1983; Ludwig and Reynold, 1988)

Environmental impact assessments ideally attempt to compare before and after studies. There were differences in sampling sites, sampling dates, effort, replication, taxonomic categories, and recovery data. (Ferson et al., 1986). Gotelli and Ellison (2004) and Odum and Barrett (2005) studied on sampling design. Diserud and Aagaard (2002) present a method that tests for changes in community structure based on repeated sampling. This may be the plot method of sampling generally consists of three major types: (1) Simple random or random sampling without replacement, (2) stratified random, and (3) systematic (Cochran, 1977)

Simple random sampling with replacement is inherently less efficient than simple random sampling without replacement (Thompson, 2002). It is important not to have to determine whether any unit in the data is included more than once. Simple random sampling consists of using a grid or a series of coordinate lines (transects) and a table of random numbers to select several plots (quadrats), the size depending on the dimensions and densities of the organisms present.

CLASSIFICATION METHODS

Measures of Similarity and Difference (Similarity and Dissimilarity)

Dissimilarity or distance measures can be categorized as metric, semi-metric, or nonmetric (McCune et al., 2002). Symmetric distances are extremely useful in community ecology. There are different Similarity coefficients based on binary data. Two methods that are simple and give good results are Jaccard and Czekanowski (Sorenson) coefficients range 0 to 1.o. Following equations are for Jaccard (1901) and Sorenson similarity coefficients

$$J = \frac{c}{a+b+c}$$

Where J= Jcacard coeffiecient, a= number of occurrences of species a alone, b= number of occurrences of b alone, c= number of co-occurences of two species (a and b).

$$C = \frac{2c}{2a+b+c}$$

Where: (same as in Jaccard)

There are different Dissimilarity coefficients based on meristic or metric data. Coefficient that are widely used in ecology studies are Bray-Curtis (1975) dissimilarity coefficient and Morisita's Index.

Bray-Curtis is recommended by Clarke and Warwick (2001) and others as the most appropriate dissimilarity coefficient for community studies.

$$BC = \frac{\sum_{j=1}^{n} |X_{1j} - X_{2j}|}{(X_{1j} + X_{2j})}$$

Where

Σ = sum (from 1 to n)
X1j = # organisms of species j (attribute) collected at site 1 (entity).
X2j = # organisms of species j collected at site 2.
BC = Bray - Curtis coefficient of distance
| | = absolute value

J = 1 to n

N = number of species

Krebs (1999) considers Morisita's index as the best similarity index, as follows:

$$C_\lambda = \frac{2\sum X_{ij}X_{ik}}{(\lambda_1 + \lambda_2)N_j N_k}$$

$$\lambda_1 = \frac{\sum [X_{ij}(X_{ij} - 1)]}{N_i(N_i - 1)}$$

$$\lambda_2 = \frac{\sum [X_{ik}(X_{ik} - 1)]}{N_k(N_k - 1)}$$

Where

Σ = sum

Cm = Morisita's Index of Similarity

Xij = No. individuals of species I in sample j

Xik = No. individuals of species I in sample k

Nj = Total No. individuals in sample j

Nk = Total No. individuals in sample k

And Coefficients of Association are two types, ranging from -1 to +1 (e.g., Pearson Correlation Coefficient) and from 0 to X (e.g., χ^2) and applicable to binary and continuous data. Pearson coefficient moment correlation uses conjoint absences, the use of which is inappropriate for comparing sites and appropriate for comparing species (Clarke and Warwick, 2001)

Euclidian Distance Euclidean distance is another measure of distance that can be applied to a site by species matrix. It has been widely used in the past because it is compatible with virtually all cluster techniques.

Cluster Analysis

Clustering is a straightforward method to show association data, however, the confidence of the nodes are highly dependent on data quality, and levels of similarity for cluster nodes is dependent on the similarity index used. Krebs

(1989) shows that mean linkage is superior to single and complete linkage methods for ecological purposes because the other two are extremes, either producing long or tight, compact clusters respectively. There are, however, no guidelines as to which mean-linkage method is the best (Swan, 1970).

The purpose of two-way clustering (also known as biclustering) is to graphically expose the relationship between cluster analyses and your individual data points. The resulting graph makes it easy to see similarities and differences between rows in the same group, rows in different groups, columns in the same group, and columns in different groups. You can see graphically how groups of rows and columns relate to each other. Two-way clustering refers to doing a cluster analysis on both the rows and columns of your matrix, followed by graphing the two dendrograms simultaneously, adjacent to a representation of your main matrix. Rows and columns of your main matrix are re-ordered to match the order of items in your dendrogram.

Table 1: Major types of hierarchical, agglomerative, polythetic clustering strategies

Group Linkage Methods

1. Nearest Neighbor
2. Farthest Neighbor
3. Median
4. Group Average
5. Centroid
6. Ward's Method
7. Flexible Beta
8. McQuitty's Method

Ward's is also know as Orloci's and Minimum Variance Method

Cluster analysis can be performed using either presence–absence or quantitative data. Each pair of sites is evaluated on the degree of similarity, and then combined sequentially into clusters to form a dendrogram with the branching point representing the measure of similarity.

TWINSPAN

The TWINSPAN method (from Two Way Indicator Species Analysis, Hill 1979; Hill et al. 1975) is a very popular method among community ecologists and it was partially inspired by the classificatory methods of classical phytosociology (use of indicators for the definition of vegetation types). Two popular agglomerative polythetic techniques are Group Average and Flexible. McCune et al. (2002) recommend Ward's method in addition. Gauch (1982) preferred to use divisive polythetic techniques such as TWINSPAN.

This method works with qualitative data only. In order not to lose the information about the species abundances, the concepts of pseudo-species and pseudo-species cut levels were introduced. Each species can be represented by several pseudo-species, depending on its quantity in the sample. A pseudo-species is present if the species quantity exceeds the corresponding cut level. TWINSPAN is a program for classifying species and samples, producing an ordered two-way table of their occurrence. The process of classification is hierarchical; samples are successively divided into categories, and species are then divided into categories on the basis of the sample classification. TWINSPAN, like DECORANA, has been widely used by ecologists.

Indicator Species Analysis

Indicator species analysis is a divisive polythetic method of numerical classification applicable to large sets of qualitative or quantitative data. This method provides a simple, intuitive solution to the problem of evaluating species associated with groups of sample units Dufrêne and Legendre's (1997). It combines information on the concentration of species abundance in a particular group and the faithfulness of occurrence of a species in a particular group. This method produces indicator values for each species in each group. These are tested for statistical significance using a Monte Carlo technique. It requires data from two or more sample units. Indicator values (range 0 for no indication to 100 for perfect indication) are presented for each species. The statistical significance of the maximum indicator value recorded for a given species is generated by a Monte Carlo test. There are many types of indicator species ranging from individual species (Dufrêne & Legendre's ,1997).

The identification of characteristic or indicator species is traditional in ecology and biogeography. Field studies describing sites or habitats usually mention one or several species that characterize each habitat. There is clearly a need for the identification of characteristic or indicator species in the fields of monitoring, conservation, and management. Indicator species add ecological meaning to groups of sites discovered by clustering, they can compare typologies derived from data analysis, identify where to stop dividing clusters into subsets, and point out the main levels in a hierarchical classification of sites.

Indicator species differ from species associations in that they are indicative of particular groups of sites. Good indicator species should be found mostly in a single group of a typology and be present at most of the sites belonging to that group (Hill, 1975). With Classical problem in community ecology and biogeography, species are the best indicators we have for particular environmental conditions. And in long-term environmental follow-up, conservation, ecological management, researchers are looking for bioindicators of habitat types to preserve or rehabilitate.

McGeoch & Chown (1998) found the indicator value method important to conservation ecosystem because it is conceptually straightforward and allows researchers to identify bioindicators for any combination of habitat types or areas of interest, e.g. existing conservation areas, or groups of sites based on the outcome of a classification procedure. Indicator species are species that, due to their niche preferences, can be used as ecological indicators of community types, habitat conditions, or environmental changes (McGeoch 1998, Carignan and Villard 2002, Niemi and McDonald 2004). They are usually determined using an analysis of the relationship between the observed species presence–absence or abundance values in a set of sampled sites and a classification of the same sites (Dufrêne and Legendre 1997).

Finally this method may represent the groups of sites in the classification, in different qualitative characteristics of the ecosystem, such as habitat or community types, environmental or succession states, or the levels of controlled experimental designs. Since indicator species analysis relates two elements, the species and the groups of sites, it can be used for gaining information on either or both. Indeed, indicator species analysis allows the characterization of the qualitative environmental preferences of the target species (for instance, when the groups are habitat types), and identifyes indicators of particular groups of sites, which can be used in further surveys.

The applications of indicator species analysis are many, including conservation, land management, landscape mapping, or design of natural reserves. Indicator species are commonly referred to as 'diagnostic species' in vegetation studies (Chytr et al. 2002).

Multi-response Permutation Procedures (MRPP)

Multi-response Permutation Procedure (MRPP) was introduced by Mielke, Berry, and Johnson (1976) as a technique for detecting the difference between a priori classified groups.

It turned out to be an extremely versatile data-analytic framework from which a number of applications fall out, such as the measurement of agreement, multivariate correlation and association coefcients, and the detection of autocorrelation (see Mielke & Berry, 2001 for a complete coverage of applications of the MRPP framework). MRPP is a non-parametric method for testing the hypothesis of no difference between two or more groups of entities. The groups must be a priori. For example, one could compare species composition between burned and unburned plots to test the hypothesis of no treatment effect. Discriminant analysis is a parametric method that can be used on the same general class of questions. However, MRPP has the advantage of not requiring assumptions (such as multivariate normality and homogeneity of variances) that are seldom met with ecological community data (Bakus, 2007).

Multiple Response Permutation Procedure (MRPP) provides a test of whether there is a significant difference between two or more groups of sampling units. This difference may be one of location (differences in mean) or one of spread (differences in within-group distance). Function MRPP operates on a data frame matrix where rows are observations and responses data matrix. The response(s) may be uni or multivariate. The method is mathematically allied with analysis of variance, in that it compares dissimilarities within and among groups. If two groups of sampling units are really different (e.g. in their species composition), then average of the within-group compositional dissimilarities ought to be less than the average of the dissimilarities between two random collection of sampling units drawn from the entire population.

The MRPP method is simply the overall weighted mean of within-group means of the pairwise dissimilarities among sampling units. The MRPP algorithm first calculates all pairwise distances in the entire dataset, then calculates delta. It then permutes the sampling units and their associated pairwise distances, and recalculates a delta based on the permuted data. It repeats the permutation step permutations times.

The function also calculates the change-corrected within-group agreement. And it also calculates classification strength which is defined as the difference between average between group dissimilarities and within group dissimilarities (Van Sickle 1997). If the first argument data can be interpreted as dissimilarities, they will be used directly. In other cases the function treats datas observations, and uses vegdist to find the dissimilarities. The default distance is Euclidean as in the traditional use of the method, but other dissimilarities in vegdist also are available.

Function meandist calculates a matrix of mean within-cluster dissimilarities (diagonal) and between-cluster dissimilarities (off-diagonal elements), and an attribute n of grouping counts. Function summary finds the within-class, between-class and overall means of these dissimilarities, and the MRPP statistics with all weight type options and the classification strength. The MRPP a robust alternative to the traditional normal theory based parametric tests, such as the t test and the analysis of variance. However, MRPP is not widely known to researchers, and part of the reason is that it has not been incorporated into major statistical packages.

Mantel Test

The Mantel test evaluates the null hypothesis of no relationship between two dissimilarity (distance) or similarity matrices. The Mantel test is an alternative to regressing distance matrices that circumvents the problem of partial dependence in these matrices.

For example evaluating the correspondence between two groups of organisms from the same set of sample units or comparing community structure before and after a disturbance is evaluate by mantel test. In Consequency, Two methods are available in PC-ORD: Mantel's asymptotic approximation and a randomization (Monte Carlo) method.

The Mantel Test (Mantel, 1967) compares two dissimilarity (distance) matrices using Pearson correlation. The matrices that use in this test must be of the same size (i.e., same number of rows). It would appear that the Mantel Test could be used to evaluate the internal structure in two sets of samples by comparing the two (dis)similarity matrices. McCune and Mefford (1999) have a computer program for performing the Mantel test. After the Mantel statistic has been calculated, the statistical significance of the relationship is tested by a permutation test or by using an asymptotic t-approximation test. The minimum number of permutations (randomizations) recommended by Manly (1997) is 1000. The Mantel and ANOSIM procedures produce similar probabilities (Legendre and Legendre, 1998).

Some of the problems associated with the Mantel test include (1) weakness in detecting spatial autocorrelation where the spatial pattern is complex and not easily modeled with distance matrices, (2) a larger number of data points may be needed for field experiments than is usually obtained, and (3) multivariate data are summarized into a single distance or dissimilarity and it is not possible to identify which variable(s) contributed the most (Fortin and Gurevitch, 2001). The use of Partial Mantel tests can distinguish the relative contributions of the factors of a third matrix considered as covariables. Mantel and partial Mantel tests can produce complementary information that other methods such ANOVA cannot provide and may do a better job than other method, in detecting block effects (Fortin and Gurevitch, 2001).

Thus, it is possible to distinguish the effects of spatial pattern from those of experimentally imposed treatment effects with these Mantel tests (Bakus, 2007).

The Best Clustering Strategies

The similarity or dissimilarity measures, forming the backbone of most multivariate clustering and ordination techniques, which are favored by many ecologists, are Jaccard and Bray–Curtis. The most popular clustering techniques in ecological studies are Group Average and Flexible group average clustering. For many of these techniques, it is impossible to choose a "best method" because of the heuristic nature of the methods (Jongman et al., 1995; Anderson, 2001).

The choice of an index must be made based on the investigator's experience, the type of data collected, and the ecological question to be answered. When comparisons are made, the Jaccard index is among the least sensitive of the similarity (or dissimilarity) indices (Jongman et al. 1995). Similarly, hierarchical agglomerative techniques (e.g., particularly Group Average) have proven to be very useful to ecologists in the construction of dendrograms. As part of a study is needed information on plant communities and their distribution. Then the objective of best method of multivariate analysis is to:
1. Identify and describe plant communities on the ecosystem
2. Map these communities to provide a tool for ecological studies and monitoring of ecological community;
3. Characterize the interrelationships between plant communities, soil particle size, moisture availability, grazing pressure and elevation.

ORDINATION METHODS

Ordination is a collective term for multivariate techniques which adapt a multi-dimensional swarm of data points in such a way that when it is projected onto a two dimensional space any intrinsic pattern the data may possess becomes apparent upon visual inspection (Pielou, 1984). As mentioned in the introduction, ordination is the arrangement of samples along gradients. Indeed, ordination can be considered a synonym for multivariate gradient analysis. Basically, ordination serves to summarize community data by producing a low-dimensional ordination space in which similar species and samples are plotted close together, and dissimilar species and samples are placed far apart.

Indirect gradient analysis
- Distance-based approaches
- Polar ordination, PO (Bray-Curtis ordination)
- Principal Coordinates Analysis, PCoA (Metric multidimensional scaling)
- Nonmetric Multidimensional Scaling, NMDS

Eigenanalysis-based approaches
- Linear model
- Principal Components Analysis, PCA
- Unimodal model
- Correspondence Analysis, CA (Reciprocal Averaging)
- Detrended Correspondence Analysis, DCA

Direct gradient analysis

- Linear model
- Redundancy Analysis, RDA
- Unimodal model
- Canonical Correspondence Analysis, CCA
- Detrended Canonical Correspondence Analysis, DCCA

Polar Ordination (Bray- Curtis)

(Polar Ordination) arranges samples with respect to "poles" (also termed end points or reference points) according to a distance matrix (Bray and Curtis 1957). These endpoints are two samples with the highest ecological distance between them (objective approach), OR two samples suspected of being at opposite ends of an important gradient (subjective approach). This procedure is especially useful for investigating ecological change (e.g., succession, recovery).

Advantages of this method is Ideal for evaluating problems with discrete endpoints. Polar Ordination ideal for testing specific hypotheses (e.g., reference condition or experimental design) by subjectively selecting the end points Disadvantages: This technique does not provide a general-purpose description of the community (perspective is biased) Very sensitive to outliers (by definition – "end points") Select a distance measure (usually Sorensen Index) and calculate matrix of distances (D) between all pairs of points (Beals, 1984). In the earliest versions of PO, these endpoints were the two samples with the highest ecological distance between them, or two samples which are suspected of being at opposite ends of an important gradient (thus introducing a degree of subjectivity).

Beals (1984) extended Bray-Curtis ordination and discussed its variants, and is thus a useful reference. The polar ordination, simplest method is to choose the pair of samples, not including the previous endpoints, with the maximum distance of separation.

Principal Component Analysis

PCA was invented in 1901 by Karl Pearson (Dunn, et al, 1987) Now it is mostly used as a tool in exploratory data analysis and for making predictive models. Principal Components Analysis is a method that reduces data dimensionality by performing a covariance analysis between factors.

It can be done by eigenvalue decomposition of a data covariance matrix or singular value decomposition of a data matrix, usually after mean centering the data for each attribute. The results of a PCA are usually discussed in terms of component scores (the transformed variable values corresponding to a particular case in the data) and loadings (the weight by which each standarized original variable should be multiplied to get the component score (Feoli and Orl¢ci. 1992). Principal components analysis is the basic eigenanalysis technique. It maximizes the variance explained by each successive axis.

PCA was one of the earliest ordination techniques applied to ecological data. PCA uses a rigid rotation to derive orthogonal axes, which maximize the variance in the data set. Both species and sample ordinations result from a single analysis. Computationally, PCA is basically an eigenanalysis. The sum of the eigenvalues will equal the sum of the variance of all variables in the data set. PCA is relatively objective and provides a reasonable but crude indication of relationships.

This method is a mathematical procedure that uses an orthogonal transformation to convert a set of observations of possibly correlated variables into a set of values of uncorrelated variables called principal components.

The number of principal components is less than or equal to the number of original variables. This transformation is defined in such a way that the first principal component has as high a variance as possible (that is, accounts for as much of the variability in the data as possible), and each succeeding component in turn has the highest variance possible under the constraint that it be orthogonal to (uncorrelated with) the preceding components.

Principal components are guaranteed to be independent only if the data set is jointly normally distributed. It is sensitive to the relative scaling of the original variables. Depending on the field of application, it is also named the discrete Karhunen–Loève transform (KLT), the Hotelling transform or proper orthogonal decomposition (POD). Broken-stick eigenvalues are provided to help evaluating statistical significance. Principal component analysis (PCA) (ter Braak and Sˇmilauer, 1998) was used to determine the association between plant communities and environmental variables, i.e. in an indirect noncanonical way (ter Braak and Loomans, 1987).

While PCA finds the mathematically optimal method (as in minimizing the squared error), it is sensitive to outliers in the data that produce large errors PCA tries to avoid. It therefore is common practice to remove outliers before computing PCA. However, in some contexts, outliers can be difficult to identify. For example in data mining algorithms like correlation clustering, the assignment of points to clusters and outliers is not known beforehand. A recently proposed generalization of PCA based on Weighted PCA increases robustness by assigning different weights to data objects based on their estimated relevancy.

Although it has severe faults with many community data sets, it is probably the best technique to use when a data set approximates multivariate normality. PCA is usually a poor method for community data, but it is the best method for many other kinds of multivariate (Bakus, 2007).

Principal Coordinate Analysis (PCoA)

Principal Coordinate Analysis (PCoA) is a method to represent on a 2 or 3 dimensional chart objects described by a square matrix containing that contains resemblance indices between these objects. This method is due to Gower (1966). It is sometimes called metric MDS (MDS: Mutidimensional scaling) as opposed to the MDS (or non-metric MDS). Both methods have the same objective and produce similar results if the similarity matrix f square distances are metric and if the dimensionality is sufficient.

Principle coordinates are similar to principal components in concept. The advantage of PCoA is that it may be used with all types of variables (Legendre and Legendre, 1998). Because of this, PCoA is an ordination method of considerable interest to ecologists. Most metric (PCoA) and nonmetric MDS plots are very similar or even identical, provided that a similar distance measure is used. The occurrence of negative eigenvalues, lack of emphasis on distance preservation, and other problems are discussed in detail by Legendre and Legendre (1998) and Clarke and Warwick (2001).

One of the biggest differences between PCA and PCoA is that the variables (i.e. species) representing the original axes are projected as biplot arrows. In the bryophyte communities, these biplot arrows greatly aid in interpretation (Bakus, 2007). In most applications of PCA (e.g. as a factor analysis technique), variables are often measured in different units. For example, PCA of taxonomic data may include measures of size, shape, color, age, numbers, and chemical concentrations. For such data, the data must be standardized to zero mean and unit variance (the typical default for most computer programs).

For ordination of ecological communities, however, all species are measured in the same units, and data should not be standardized. In matrix algebra terms, most PCAs are eigenanalyses of the correlation matrix, but for ordination they should be PCAs of the covariance matrix. In contrast to Correspondence Analysis and related methods, species are represented by arrows. This implies that the abundance of the species is continuously increasing in the direction of the arrow, and decreasing in the opposite direction. Thus PCA is a 'linear method'.

Although the discussion above implies that PCA is distinctly different from PCoA, the two techniques end up being identical, if the distance metric is Euclidean. Unfortunately, this linear assumption causes PCA to suffer from a serious problem, the horseshoe effect, which makes it unsuitable for most ecological data sets (Gauch 1982). Principal Coordinates Analysis (PCoA, = Multidimensional scaling, MDS) is a method to explore and to visualize imilarities or dissimilarities of data. It starts with a similarity matrix or dissimilarity matrix (= distance matrix) and assigns for each item a location in a lowdimensional space, e.g. as a 3D graphics. PCOA tries to find the main axes through a matrix. It is a kind of eigenanalysis (sometimes referred as "singular value decomposition") and calculates a series of eigenvalues and eigenvectors. Each eigenvalue has an eigenvector, and there are as many eigenvectors and eigenvalues as there are rows in the initial matrix. Eigenvalues are usually ranked from the greatest to the least. The first eigenvalue is often called the "dominant" or leading" eigenvalue. Using the eigenvectors we can visualize the main axes through the initial distance matrix. Eigenvalues are also often called "latent values". The result is a rotation of the data matrix: it does not change the positions of points relative to each other but it just changes the coordinate systems!

By using PCoA we can visualize individual and/or group differences. Individual differences can be used to show outliers. There is also a method called 'Principal Component Analysis' (PCA, sometimes also misleadingly abbreviated as 'PCoA') which is different from PCOA.

- Principal Coordinates Analysis (Principal Coordinates Analysis (PCoA PCoA or PCO) or PCO)
- Maximizes the linear correlation between distance measures and distance in the ordination.
- useful if one has only a distance (or similarity) matrix
- The underlying model is that there a fixed number of explanatoryoriginal variables. In contrast, PCA, RA, and DCA assume that there are potentially many variables, but of declining importance.
- One cannot easily put new points in a PCoA.
- For Euclidean distance, PCoA= PCA
- PCoA can be expressed as an eigenanalysis.

Factor Analysis (FA)

FA and PCA (principal components analysis) are methods of data reduction. Take many variables and explain them with a few "factors" or "components". Correlated variables are grouped together and separated from other variables with low or no correlation. Patterns of correlations are identified and either used as descriptives (PCA) or as indicative of underlying theory (FA). It process of providing an operational definition for latent construct (through regression equation). FA and PCA are not much different than canonical correlation in terms of generating canonical variates from linear combinations of variables. Although there are now no "sides" of the equation.

For calcuting this method we should:
1. Selecting and measuring a set of variables in a given domain
2. Data screening in order to prepare the correlation matrix
3. Factor Extraction
4. Factor Rotation to increase interpretability
5. Interpretation

Factor analysis is seldom used in ecology. Several statisticians state that it should not be used because it is based on a special statistical model. Estimating the unique variance is the most difficult and ambiguous task in FA (McGarigal et al., 2000).

Redundancy Analysis

RDA is a linear method and since it is a linear method, species as well as environmental variables are represented by arrows. In most cases, it is best to represent the two sets of arrows in two figures for ease of display. Thus, if have a gradient along which all species are positively correlated, RDA will detect such a gradient while CCA will not. RDA can use 'species' that are measured in different units. If so, the data must be centered and standardized. But in fact, as an ordination technique, the species should not be standardized. Redundancy analysis is the linear method of direct ordination. It also goes under the name of principal components analysis with respect to instrumental variables, which in our case are environmental variables (Sabatier et al., 1989). It is also called least-squares reduced rank regression so as to emphasize its link with multivariate regression (ter Braak & Prentice, 1988; ter Braak & Looman, 1994). In statistical textbooks on multivariate analysis, redundancy analysis is usually neglected.

RDA is useful when gradients are short. In particular, RDA may be the method of choice in a short-term experimental study. In such cases, the treatments are the explanatory variables (and are usually dummy variables). The sample ID or block might be a covariable in a partial RDA, if one wishes to factor out local effects (Bakus, 2007). 'variance explained' is actually a variance explained, and not merely inertia. Thus, variance partitioning, and interpretation of eigenvalues, are more straightforward than for CCA (Lepš & Šmilauer, 1999).

The relatively short space devoted to RDA should not be given as an indication that it is less valuable than, or inferior to, other method. It is simply used for different purposes. Consequently, the linear method of direct ordination, redundancy analysis, can often efficiently display the interesting effects, and there is no need for methods that also work when the range of community variation is larger (such as canonical correspondence analysis) nor for unconstrained nonmetric multidimensional scaling.

Correspondence Analysis (CA) or Reciprocal Averaging (RA)

Reciprocal averaging is also known as correspondence analysis (CA) because one algorithm for finding the solution involves the repeated averaging of sample scores and species scores (citations). It is a graphical display ordination technique which simultaneously displays the rows (sites) and columns (species) of a data matrix in low dimensional space (Gittins, 1985). Row identifiers (species) plotted close together are similar in their relative profiles, and column identifiers plotted close together are correlated, enabling one to interpret not only which of the taxa are clustered, but also why they are clustered (Bakus,2007).

Reciprocal averaging (RA) yields both normal and transpose ordinations automatically. Like DCA, RA ordinates both species and samples simultaneously. Instead of maximizing 'variance explained', CA maximizes the correspondence between species scores and sample scores. If species scores are standardized to zero mean and unit variance, the eigenvalues also represent the variance in the sample scores (but not, as is often misunderstood, the variance in species abundance).

Since CA is a unimodal model, species are represented by a point rather than an arrow. This is (under some choices of scaling; see ter Braak and Šmilauer 1998) the weighted average of the samples in which that species occurs. With some simplifying assumptions (ter Braak and Looman 1986), the species score can be considered an estimate of the location of the peak of the species response curve.

The CA distortion is called the arch effect, which is not as serious as the horseshoe effect of PCA because the ends of the gradients are not incurved. Nevertheless, the distortion is prominent enough to seriously impair ecological interpretation (Bakus, 2007). In other words, the spacing of samples along an axis may not affect true differences in species composition. Gradient compression can be quite blatant in simulated data sets. The problems of gradient compression and the arch effect led to the development of Detrended Correspondence Analysis (Bakus, 2007).

Detrended Correspondence Analysis (DCA)

Detrended Correspondence Analysis (DCA) eliminates the arch effect by detrending (Hill and Gauch 1982). It's a series of rules that are used to reshape data to make it friendlier for analysis. Once again, primarily used for ecological data, but can be extended to anything (data simply can't contain negative values). The reason that this technique is used is to overcome the arch effect (the horseshoe effect). Found in data whenever "PCA or other distance conserving ordination techniques are applied to data which follow a continuous gradient, along which there is a progressive turnover of dominant variables." Such as in ecological succession.

After ordination by a distance conserving technique and the first two axes are plotted against each other, one would find an arch shape. DCA is another eigenanalysis ordination technique that based on reciprocal averaging (RA; Hill 1979). DCA is geared to ecological sets, is based on samples and species. DCA ordinates both species and samples simultaneously. There are two basic approaches to detrending: by polynomials and by segments (ter Braak and Šmilauer 1998). Detrending by polynomials is the more elegant of the two: a regression is performed in which the second axis is a polynomial function of the first axis, after which the second axis is replaced by the residuals from this regression. Similar procedures are followed for the third and higher axes.

The compression of the ends of the gradients is corrected by nonlinear rescaling. Rescaling shifts sample scores along each axis such that the average width is equal to 1. Rescaling has a beneficial consequence: the axes are scaled in units of beta diversity (SD units, or units of species standard deviations). Thus if the underlying gradient is important well known, it is possible to plot the DCA scores as a function of the gradient, and there by determine whether the species 'perceive' the gradient differently than we measure it. The shape of the species response curves may change if axes are rescaled. Thus, skewness and kurtosis are largely artifacts of the units of measurement for which we choose to measure the environment.

Nonmetric multimentional scaling (MDS, NMDS, NMS, NM-MDS)

Nonmetric Multidimensional Scaling (NMDS) rectifies this by maximizing the rank order correlation. For this proceeds at first the user selects the number of dimensions (N) for the solution, and chooses an appropriate distance metric and then The distance matrix is calculated. And initial configuration of samples in N dimensions is selected. This configuration can be random, though the chances of reaching the correct solution are enhanced if the configuration is derived from another ordination method. And finally, the final configuration of points represents your ordination solution. The configuration is dependent on the number of dimensions selected; e.g. the first two axes of a 3-dimensional solution does not necessarily resemble a 2-dimensional solution. The stress will typically decrease as a function of the number of dimensions chosen; this function can aid in the selection of the results (Bakus, 2007).

This is why it is sometimes useful to rotate the solution (such as by the Varimax method) – although there is no theory that states that the final solution will represent a 'gradient' Other problems and advantages of NMDS will be discussed later, when comparing it to Detrended Correspondence Analysis (Bakus, 2007).

MANOVA AND MANCOVA

A factorial MANOVA may be used to determine whether or not two or more categorical grouping variables (and their interactions) significantly affect optimally weighted linear combinations of two or more normally distributed outcome variables. These parametric multivariate techniques (Multivariate Analysis of Variance and Multivariate Analysis of Covariance) are similar to ANOVA and ANCOVA MANOVA (Wilks' Lambda) and ANCOVA are advantageous in that performing multiple univariate tests can inflate the a value, leading to false conclusions (Scheiner, 2001).

MANOVA seeks differences in the dependent variables among the groups (McCune et al, 2002). Assumptions of MANOVA include multivariate normality (error effects included), independent observations, and equality of variance-covariance matrices (Paukert and Wittig, 2002). Because of these assumptions, among others, MANOVA is not often used in ecology although its use in increasing. The power of traditional MANOVA declines with an increase in the number of response variables (Scheiner, 2001). Unequal sample sizes are not a large problem for MANOVA, but may bias the results for factorial or nested designs. Before ANCOVA is run, tests of the assumption of homogeneity of slopes need to be performed (Petratis et al., 2001).

Early attempts to develop nonparametric multivariate analysis include those of Mantel and Valand (1970). They are more complex as they handle three or more variables simultaneously. They are frequently used with the analysis of experimental studies, especially laboratory experiments. More recently, the Analysis of Similarities was developed to compare communities or changes in communities because of pollution (Clarke, 1993). For typical species abundance matrices, an Analysis of Similarities (ANOSIM) permutation procedure is recommended over MANOVA.

Multiple analysis of variance (MANOVA) is used to see the main and interaction effects of categorical variables on multiple dependent interval variables. MANOVA uses one or more categorical independents as predictors, like ANOVA, but unlike ANOVA, there is more than one dependent variable. Where ANOVA tests the differences in means of the interval dependent for various categories of the independent(s), MANOVA tests the differences in the centroid (vector) of means of the multiple interval dependents, for various categories of the independent(s). One may also perform planned comparison or post hoc comparisons to see which values of a factor contribute most to the explanation of the dependents. There are multiple potential purposes for MANOVA.

To compare groups formed by categorical independent variables on group differences in a set of interval dependent variables. To use lack of difference for a set of dependent variables as a criterion for reducing a set of independent variables to a smaller, more easily modeled number of variables. Multiple analysis of covariance (MANCOVA) is similar to MANOVA, but interval independents may be added as "covariates." These covariates serve as control variables for the independent factors, serving to reduce the error term in the model. Like other control procedures, MANCOVA can be seen as a form of "what if" analysis, asking what would happen if all cases scored equally on the covariates, so that the effect of the factors over and beyond the covariates can be isolated. The discussion of concepts in the ANOVA section also applies, including the discussion of assumptions.

Discriminate Analysis

Discriminate Analysis (DA) is a powerful tool that can be used with both clusters of species data and environmental variables. It determines which variables discriminate between two or more groups, that is, independent variables are used as predictors of group membership (McCune et al., 2002). It is very similar to MANOVA and multiple regression analysis (Statsoft, Inc., 1995; McGarigal et al., 2000). Clusters can be identified by several methods using raw data: (1) constructing a dendrogram, (2) using PCA (even if you have field data) for initial visual identification of clusters, and (3) point rotation in space by rotating ordinations (i.e., rotating axes – see McCune and Mefford, 1999). If any method indicates groups or clusters of data then DA can be used. However, the number of groups is set before the DA analysis. DA finds a transform for the minimum ratio of difference between pairs of multivariate means and variances in which two clusters are separated the most and inflated the least. DA produces two functions: (1) classification function consisting of 2 groups or clusters of points (this information can be used for prediction with probabilities) and (2) discriminate function containing environmental variables that can be used to discriminate differences among the groups.

DA differs from PCA and Factor Analysis in that no standardization of data is needed (PCA and FA need standardization because of scaling problems) and the position of the axes distinguishes the maximum distance between clusters. (Davis, 1986). DA assumes a multivariate normal distribution, homogeneity of variances, and independent samples (Paukert and Wittig, 2002). Violations of normality are usually not fatal (i.e., somewhat non-normal data can be used). A description of the procedure to use DA with Statistica is given in the Appendix. Multiple Discriminate Analysis (MDA) is the term often used when three or more clusters of data are processed simultaneously. MDA is particularly susceptible to rounding error. Calculations in double precision for at least the eigenvalueeigenvector routines are advisable (Green, 1979). Limitations of Discriminate Analysis are discussed by McGarigal et al. (2000) Dytham (1999).

Canonical Correspondence Analysis (CCA)

In ecology studies, the ordination of samples and species is constrained by their relationships to environmental variables. When species responses are unimodal (humpshaped), and by measuring the important underlying environmental variables, CCA is most likely to be useful. It was used to examine the relationships between the measured variables and the distribution of plant communities (Ter Braak, 1986). CCA expresses species relationships as linear combinations of environmental variables and combines the features of CA with canonical correlation analysis (Green, 1989). This provides a graphical representation of the relationships between species and environmental factors.

Canonical Correlation Analysis is presented as the standard method to relate two sets of variables (Gittins, 1985). However, the latter method is useless if there are many species compared to sites, as in many ecological studies, because its ordination axes are very unstable in such cases.

The best weight for CCA describes environment variables with the first axis shows. Species information structure using a reply CCA Nonlinear with the linear combination of variables will consider environmental characteristics of acceptable behavior characteristics of species with environment shows. CCA analysis combined with non-linear species and environmental factors shows the most important environmental variable in connection with the axes shows. In Canonical Correspondence Analysis, the sample scores are constrained to be linear combinations of explanatory variables. CCA focuses more on species composition, i.e. relative abundance.

When a combination of environmental variables is highly related to species composition, this method, will create an axis from these variables that makes the species response curves most distinct. The second and higher axes will also maximize the dispersion of species, subject to the constraints that these higher axes are linear combinations of the explanatory variables, and that they are orthogonal to all previous axis. Monte Carlo permutation tests were subsequently used within canonical correspondence analysis (CCA) to determine the significance of relations between species composition and environmental variables (ter Braak, 1987)

The outcome of CCA is highly dependent on the scaling of the explanatory variables. Unfortunately, we cannot know a priori what the best transformation of the data will be, and it would be arrogant to assume that our measurement scale is the same scale used by plants and animals. Nevertheless, we must make intelligent guesses (Bakus, 2007). In CCA possible that patterns result from the combination of several explanatory variables. And many extensions of multiple regression (e.g. stepwise analysis and partial analysis) also apply to CCA. It is possible to test hypotheses (though in CCA, hypothesis testing is based on randomization procedures rather than distributional assumptions). Explanatory variables can be of many types (e.g. continuous, ratio scale, nominal) and do not need to meet distributional assumptions.

Another advantage of CCA lies in the intuitive nature of its ordination diagram, or triplot. It is called a triplot because it simultaneously displays three pieces of information: samples as points, species as points, and environmental variables as arrows (or points). If data sets are few, CCA triplots can get very crowded then should be separate the parts of the triplot into biplots or scatterplots (e.g. plotting the arrows in a different panel of the same figure) or rescaling the arrows so that the species and sample scores are more spread out. And we can only plotting the most abundant species (but by all means, keep the rare species in the analysis).

Noise in the species abundance data set is not much of a problem for CCA (Palmer, 1988). However, it has been argued that noise in the environmental data can be a problem (McCune 1999). It is not at all surprising that noise in the predictor variables will cause noise in the sample scores, since the latter are linear combinations of the former. It is probably obvious that the choice of variables in CCA is crucial for the output. Meaningless variables will produce meaningless results. However, a meaningful variable that is not necessarily related to the most important gradient may still yield meaningful results (Palmer, 1988).

If many variables are included in an analysis, much of the inertia becomes 'explained'. Any linear transformation of variables (e.g. kilograms to grams, meters to inches, Fahrenheit to Centigrade) will not affect the outcome of CCA whatsoever. There are as many constrained axes as there are explanatory variables. The total 'explained inertia' is the sum of the eigenvalues of the constrained axes. The remaining axes are unconstrained, and can be considered 'residual'. The total inertia in the species data is the sum of eigenvalues of the constrained and the unconstrained axes, and is equivalent to the sum of eigenvalues, or total inertia, of CA. Thus, explained inertia, compared to total inertia, can be used as a measure of how well species composition is explained by the variables. Unfortunately, a strict measure of 'goodness of fit' for CCA is elusive, because the arch effect itself has some inertia associated with it (Bakus, 2007).

The ordination diagrams of canonical correlation analysis and redundancy analysis display the same data tables; the difference lies in the precise weighing of the species (ter Braak & Looman, 1994; Van der Myer, 1991) One of limitations to CCA is that correlation does not imply causation, and a variable that appears to be strong may merely be related to an unmeasured but 'true' gradient. As with any technique, results should be interpreted in light of these limitations (McCune, 1999).

Multiple Regression (MR) (Multiple Linear Regression)

The mechanics of testing the "significance" of a multiple regression model is basically the same as testing the significance of a simple regression model, we will consider an F-test, a ttest (multiple t's) and R-sqrd. However, unlike simple regression where the F & t tests tested the same hypothesis, in multiple regression these two tests have different purposes. R-sqrd is still the percent of variance explained but is no longer the correlation squared (as it was with in simple linear regression) and we will also introduce adjusted R-sqrd. When considering a multiple regression (MR) model the most common order to interpret things consists of first looking at the R-sqrd, then testing the entire model by looking at the F-test, and finally looking at each individual coefficient individually using the t-tests. NOTE: The term "significance" is a nice convenience but is very ambiguous in definition if not properly specified. Thus when taking this class you should avoid simply saying something is significant without explaining (1) how you made that determination, and (2) what that specifically means in this case. You will see from the examples that those two things are always done. If you cannot do that then any time you use the word "significant" you are potentially hurting yourself in two ways; (1) you won't do well on the quizzes or exams where you have to be able to be more explicit than simply throwing out the word "significant", and (2) you will look like a fool in the business world when somebody asks you to explain what you mean by "significant" and you are stumped. Remember if you can't explain your results in managerial terms than you do not really understand what you are doing.

In order to show the relationship between biotic (principal component axes) and abiotic factors (environmental factors), a multiple regression type analysis is used. Multiple regression solves simultaneously normal equations and produces partial regression coefficients. Partial regression coefficients each give the rate of change or slope in the dependent variable for a unit of change in a particular independent variable, assuming all other independent variables are held constant. MR is not considered by some statisticians as a multivariate procedure because it includes only one dependent variable (Paukert and Wittig, 2002).

The objective of multiple regression is to determine the influence of independent variables on a dependent variable, for example, the effect of depth, sediment grain size, salinity, temperature, and predator density on the population density of species. The parameters are estimated by the least-squares method that is, minimizing the sum of squares of the differences between the observed and expected response (Jongman et al., 1995)

Normally component loadings (e.g., scores in PCA) suggest which variables are most important. However, with species abundances the only variable in some ordinations, one must use other techniques to attempt to suggest what may have produced the gradients for Axis 1, 2, 3, and so forth. Univariate analyses such as the Spearman rank correlation coefficient are not as ecologically realistic as multivariate analyses such as multiple regression because some variables are correlated and there are interaction effects between variables (Jongman et al., 1995). The highest standardized partial regression coefficients (positive or negative) suggest the most important factors (e.g., sediment size, predator density, etc.) in controlling the population density of species X. Significance tests and standard errors then can be calculated from the data.

Multiple regression has many potential problems such as the type of response curve, error distribution, and outliers that may unduly influence the results (Jongman et al., 1995). Multiple regression variables may be highly correlated, therefore, examine the correlation coefficients first (i.e., run a multiple correlation analysis between variables) to exclude some of them before doing multiple regressions. Multiple regression generally should employ a maximum of 6 variables. Legendre and Legendre (1998) suggest a stepwise procedure for reducing numerous variables. This involves a process of alternating between forward selection and backward elimination (Kutner et al., 1996). Lee and Sampson (2000) took ordination scores, representing a gradient in fish communities, and regressed them against a group of environmental variables and time. Many of scientist study on multiple regression and computer program (Such as Davis, 1986; and Sokal and Rohlf, 1995)

In multiple regression, it is typical to include quadratic terms for explanatory variables. For example, if you expect a response variable to reach a maximum at an intermediate value of an explanatory variable, including this explanatory variable AND the square of the explanatory variable may allow a concave-down parabola to provide a reasonable fit. This is an analogous situation to multiple regression: the multiple r2 or 'variance explained' increases as a function of the number of variables included. Both multiple regression and CCA find the best linear combination of explanatory variables, they are not guaranteed to find the true underlying gradient (which may be related to unmeasured or unmeasurable factors), nor are they guaranteed to explain a large portion of variation in the data.

Path Analysis

In statistics, path analysis is used to describe the directed dependencies among a set of variables. This includes models equivalent to any form of multiple regression analysis, factor analysis, canonical correlation analysis, discriminate analysis, as well as more general families of models in the multivariate analysis of variance and covariance analyses (MANOVA, ANOVA, ANCOVA).

Path analysis is a straightforward extension of multiple regression. Its aim is to provide estimates of the magnitude and significance of hypothesised causal connections between sets of variables. This is best explained by considering a path diagram. Path analysis is an extension of multiple linear regression, allowing interpretation of linear relationships among a small number of descriptors (Legendre and Legendre, 1998). This method was originally developed by Sewall Wright in which he introduced the concept of a path diagram. It handles more than one dependent variable and the effects of dependent variables on one another (Mitchell, 2001). Path analysis assumes a normal distribution of residuals, additive and linear effects, inclusion of all important variables, that residual errors are uncorrelated, and that there is no measurement error.

Path analysis was developed as a method of decomposing correlations into different pieces for interpretation of effects (e.g., how does parental education influence children's income 40 years later?). Path analysis is closely related to multiple regression; you might say that regression is a special case of path analysis. Some people call this stuff (path analysis and related techniques) causal modeling. The reason for this name is that the techniques allow us to test theoretical propositions about cause and effect without manipulating variables.

However, the causal in causal modeling refers to an assumption of the model rather than a property of the output or consequence of the technique. That is, people assume some variables are causally related, and test propositions about them using the techniques.

Canonical Correlation Analysis (CVA)

Canonical variate analysis (CVA) is a widely used method for analyzing group structure in multivariate data. It is mathematically equivalent to a one-way multivariate analysis of variance and often goes by the name of canonical discriminate analysis. Change over time is a central feature of many phenomena of interest to researchers. This dissertation extends CVA to longitudinal data. It develops models whose purpose is to determine what is changing and what is not changing in the group structure. Three approaches are taken: a maximum likelihood approach, a least squares approach, and a covariance structure analysis approach. All methods have in common that they hypothesize canonical variates which are stable over time.

The maximum likelihood approach models the positions of the group means in the subspace of the canonical variates. It also requires modeling the structure of the within-groups covariance matrix, which is assumed to be constant or proportional over time. In addition to hypothesizing stable variates over time, one can also hypothesize canonical variates that change over time. Hypothesis tests and confidence intervals are developed. The least squares methods are exploratory. They are based on three-mode PCA methods such as the Tucker2 and parallel factor analysis. Graphical methods are developed to display the relationships between the variables over time.

Stable variates over time imply a particular structure for the between-groups covariance matrix. This structure is modeled using covariance structure analysis, which is available in the SAS package Proc Calis. Canonical Variate Analysis is a special case of Canonical Correlation Analysis (Jongman et al., 1995). It is also described as a type of linear discriminate analysis (McGarigal et al., 2000). The set of environmental variables consists of a single nominal variable defining the classes.

CVA is usable only if the number of sites is much greater than the number of species and the number of classes. Many ecological data sets cannot be analyzed by CVA without dropping many species, thus CVA is not used much in ecology. Instead, they usually give the impression that there is only one such hypothesis, and therefore only one statistical technique is needed-Hotelling's canonical variate analysis (CVA). Most discussions of CVA are restricted almost entirely to a description of the underlying mathematical theory, computing directions, and perhaps an example of the computations.

Very little is usually said about the logic of the method, so that the reader is unable to judge for himself whether the method described can actually be used to test the hypothesis of interest to him.

Even when CVA is appropriate, several prominent sources have recommended misleading interpretations of the statistics computed in CVA. Perhaps because of these deficiencies, many behavioral scientists have concluded incorrectly that CVA has few or no valid and important uses in the behavioral sciences. The use originally proposed by Hotelling has been rejected by most behavioral scientists.

ORDINATION AND CLASSIFICATION METHODS IN VARIOUS ECOLOGY STUDIES

The more applied an ecological study is, the more the emphasis is on the effects on ecological communities of particular environmental factors, for example pollutants, management regimes, and other human-induced changes in the environments. (ter Braak, 1994). Correspondingly, the statistical analysis should not 'just' show the major variation in the species assemblage, but focus on the effects on the variables of prime interest. Applied studies thus call for direct methods of ordination, typically with a very limited number of (qualitative or quantitative) environmental variables (ter Braak, 1994) The range of community variation in an applied study tends to be quite small compared to that in the early ordination studies (e.g. Whittaker, 1965; Hill & Gauch, 1980).

Determining which factors control the distribution patterns of plant communities remains a central goal in ecology Classification assumes from the outset that the species assemblages fall into discontinuous groups, whereas ordination starts from the idea that such assemblages vary gradually. Ordination compares sites on their degree of similarity, and then plots them in Euclidian space, with the distance between points representing their degree of Similarity. Ordination techniques include principal components analysis (PCA), detrended correspondence analyses (DCA), and nonmetric multidimensional scaling (NMS)

Ordination (or inertia) methods, like principal component and correspondence analysis, and clustering and classification methods are currently used in many ecological studies (Zare Chahouki et al., 2009, Anderson, 2002; Gauch et al., 1977; Orloci, 1975; Whittaker (ed.), 1967; Legendre & Legendre, 1988). Ordination methods can be divided in two main groups, direct and indirect methods. Direct methods use species and environment data in a single, integrated analysis. Indirect methods use the species data only (Jongman et al.,

1987). In contrast, if a unimodal response model is assumed, the relationships are unimodal. Unimodal relations are usefully summarized by their modes or - more conveniently - weighted averages (ter Braak & Looman, 1986), so that a sensible coefficient for the species × environment table is the weighted average. The other way round, if a correlation coefficient is chosen, the implied response model is linear or approximately linear and if the chosen coefficient is the weighted average than the implied response model is unimodal (i.e. if the true model is bimodal, the ordination will fail, and if the true model is linear, the ordination will be inefficient). Assumptions about the response model and the choice of coefficients to use in secondary tables are thus interrelated.

New methods of exploring differences among groups include the nonparametric, recursive classification, and regression tree (CART). It is used to classify habitats or vegetation types and their environmental variables (McCune et al., 2002). It produces a top-to-bottom visual classification tree that undergoes a "pruning" or optimization process. CART is used to generate community maps, wildlife habitats, and land cover types in conjunction with a GIS (see p. 195 in Chapter 3). Another multivariate technique is Structural Equation Modeling (unfortunately termed SEM), a merger of factor analysis and path analysis (McCune et al. (2002). It is a method of evaluating complex hypotheses (e.g., effects of abiotic factors on plant species richness).

With multiple causal pathways among variables. It requires the initial development of a path diagram. It is an analysis of covariance relationships, effectively limited to about 10 variables. See Shipley (2000), McCune et al. (2002), Pugesek et al. (2002) and Bakus (2007).

REFERENCES

1. Anderson, M.J. (2001). A new method for non-parametric multivariate analysis of variance. Austral. Ecol. 26:32-46.
2. Alisauskas, R.T. (1998). Winter range expansion and relationships between landscape and morphometrics of midcontinent Lesser Snow Geese. Auk 115(4):851-862.
3. Anderson, C.W., Barnett, V., Chatwin, P.C. and El-Shaarawi A.H. (2002). Quantitative Methods for Current EnvironmentalIssues. Springer-Verlag, New York.
4. C.T. Bastian, S.R. Koontz and Menkhaus D.J. (2001). The Impact of Forward Contract Information on the Fed Cattle Market: An Experimental Investigation into Mandatory Price Reporting, UW Agricultural and Applied Economics Seminar. August 31, 2001. (Presented by Bastian).
5. Bakus Gerald J. (2007). Quantitative Analysis of Marine Biological

Communities Field Biology and Environment. WILEY-INTERSCIENCE, A John Wiley & Sons, Inc., Publication, 453p.
6. Beals, E.W. (1973). Ordination: Mathematical elegance and ecological naivete. J. Ecol. 61:23– 35.
7. Bray, J.R. and Curtis J.T. (1957). An ordination of the upland forest communities of southern Wisconsin.Ecol. Monogr. 27:325-349.
8. Carignan, V. and Villard M. (2002). Selecting indicator species to monitor ecological integrity: a review. Environ. Monitor. Assess. 78: 45-61.
9. Chytr, M. (2002). Determination of diagnostic species with statistical fidelity measures. J. Veg. Sci. 13: 79–90.
10. Clarke, K.R. (1993). Non-parametric multivariate analyses of changes in community structure. Aust. J. Ecol.18:117-143.
11. Clarke, K.R. and Warwick R.M. (2001). Change in Marine Communities: An Approach to Statistical Analysis and Interpretation. 2nd edition. PRIMER- E, Plymouth Marine Laboratory, Plymouth, U.K.
12. Clifford, H.T. and Stephenson W. (1975). An introduction to numerical classification. Academic Press, New York, pp. 229.
13. Cochran, W.G. (1977). Sampling Techniques. Wiley, New York.
14. Davis, J.C. (1986). Statistics and Data Analysis in Geology. Wiley, New York.
15. Diserud, O.H. and Aagaard K. (2002). Testing for changes in community structure based on repeated sampling. Ecology, 83(8): 2271–2277.
16. Dufrene, M. and Legendre P. (1997). Species assemblages and indicator species: the need for a flexible asymmetrical approach. Ecol. Monogr. 67:345–366.
17. Dunn, C. P., and Stearns F. (1987). Relationship of vegetation layers to soils in southeastern Wisconsin forested wetlands. Am. Midl. Nat. 118:366-74.
18. Dytham, C. (1999). Choosing and Using Statistics: A Biologist's Guide. Blackwell Publishing, Williston, VT.
19. Ferson, S., Downey P., Klerks P., Weissburg M., Kroot S.I., S. Jacquez O., Ssemakula J., Malenky R. and Anderson K. (1986). Competing reviews, or why do Connell and Schoener disagree? Am. Nat., 127: 571–576.
20. Feoli, E., and Orl¢ci L. (1992). Thre properties and interpretation of observations in vegetation study. Coenoses 6:61-70.
21. Fortin, M-J. and Gurevitch J. (2001). Mantel tests: Spatial structure in field experiments. pp. 308-326 in: Scheiner S.M. and J. Gurevitch (eds.).

Design and Analysis of Ecological Experiments. Oxford University Press, Oxford.
22. Gauch, H.G. (1977). ORDIFLIX—A flexible computer program for four ordination techniques: weighted averages, polar ordination, principal components analysis, and reciprocal averaging. In: Ecology and Systematics, Cornell University, Ithaca, N.Y.
23. Gauch, H.G., J. (1982). Multivariate Analysis in Community Ecology. Cambridge University Press, New York.
24. Gittins, R. (1985). Canonical analysis: a review with applications in ecology. SpringerVerlag, Berlin.
25. Gotelli, N.J. and Ellison A.M. (2004). A Primer of Ecological Statistics. Sinauer Associates, Sunderland, Maine
26. Gower, J. C. (1966). Some Distance Properties of Latent Root and Vector Methods used in Multivariate Analysis. Biometrika 53, 325-338.
27. Green, R.H. (1979). Sampling Design and Statistical Methods for Environmental Biologists. John Wiley and Sons, New York.
28. Greig-Smith, P. (1983). Quantitative Plant Ecology, 3rd Edition. Blackwell Scientific Publications, London, 359 pp.
29. Green, R.H. 1989. Power analysis and practical strategies for environmental monitoring. Environ. Res. 50:195–205.
30. Hill, M.O., Bunce R.G.H. & Shaw M.V. (1975). Indicator species analysis, a divisive polythetic method of classification, and its application to survey of native pinewoods in Scotland. Journal of Ecology, 63: 597–613
31. Hill, M.O. (1979). TWINSPAN – a FORTRAN Program for Arranging Multivariate Data in an Ordered Two-way Table by Classification of the Individuals and Attributes. Ithaca: Section of Ecology and Systematic, Cornell University.
32. Hill M.O. & Gauch H.G. (1980). Detrended correspondence analysiss, an improved ordination technique. Vegetatio, 42: 47-58
33. Jaccard, P. (1901): Etude comparative de la distribution florale dans une portion des Alpes et du Jura. Bulletin de la Socie´ te´ Vaudoisedes SciencesNaturelles, 37: 547–579.
34. Jongman, R.H.G., ter Braak C.J.F. and Van Tongeren O.F.R. (1987). Data analysis in community and landscape ecology. Cambridge University Press, Cambridge, UK.
35. Jongman, R.H.G., Ter Braak, C.J.F. and van Tongeren O.F.R. (eds.) (1995). Data Analysis in Community and Landscape Ecology. Cambridge University Press, Cambridge.

36. Kent, M. (2006). Numerical classification and ordination methods in biogeography. Progress in Physical Geography, 30: 399-408
37. Krebs, C.J. (1989). Ecol. Method. Harper Collins, New York.
38. Krebs, C.J. (1999). Ecological Methodology. Harper & Row, New York.
39. Kutner, M.H., Nachtscheim, C.J., Wasserman, W. and Neter J. (1996). Applied linear statistical models. WCB/McGraw-Hill, New York.
40. Lee, Y.W. and Sampson D.B. (2000). Spatial and temporal stability of commercial ground fish assemblages off Oregon andWashington as inferred from Oregon travel logbooks. Canadian J. Fish. Aqua. Sci., 57:2443-2454.
41. Legendre, P. & Gallagher E.D. (2001). Ecologically meaningful transformations for ordination of species data. Oecologia, 129: 271–280
42. Legendre, P. and Legendre L. (1998). Numerical Ecology. 2nd Edition. Elsevier, Amsterdam.
43. Lepš, Jan & Šmilauer P. (1999). Multivariate Analysis of Ecological Data Faculty of Biological Sciences, University of South Bohemia Ceské Budejovice,110pp
44. Ludwig, J.A., Reynold, J.F. (1988). Statistical Ecology. Wiley, New York, 337pp
45. Niemi, G.J. and McDonald M.E. (2004). Application of ecological indicators. – Annu. Rev. Ecol. Evol. Syst. 35: 89–111.
46. Manly, B.F.G. (1997). Randomization, Bootstrap and Monte Carlo Methods in Biology. Chapman and Hall, London.
47. Mantel, N. 1967. The detection of disease clustering and generalized regression approach. Cancer Res. 27:209-220.
48. Mantel, N. and Valand R.S. (1970). A technique of nonparametric multivariate analysis. Biometrics 26:547–558.
49. McCune, B. and Mefford M.J. (1999). PCORD, Multivariate Analysis of Ecological Data, Version 4. MjM Software Design, Gleneden Beach, Oregon, USA.
50. McCune, B., Grace J.B. and Urban D.L. (2002). Analysis of Ecological Communities. MjM Software Design, Gleneden Beach, Oregon.
51. McGarigal, K., Cushman, S. and Stafford S. (2000). Multivariate Statistics for Wildlife and Ecology Research. Springer-Verlag, New York
52. McGeogh, M.A. (1998). The selection, testing and application of terrestrial insects as bioindicators. Biol. Rev. 73: 181–201.
53. McGeoch, M.A. and Chown. S.L. (1998). Scaling up the value of

bioindicators. Trends Ecol. Evol. 13: 46-47.
54. Mielke, P.W. & Berry K.J. (2001). Permutation methods: A distance function approach. New York: Springer-Verlag.
55. Mielke, P. W., Berry K.J. & Johnson E. S. (1976). Multi-response permutation procedures for a priori classications. Communications in Statistics- Theory and Methods, 5: 1409- 1424
56. Mitchell, R.J. (2001). Path analysis: Pollination. pp. 217–234 in: Scheiner S.M. and J. Gurevitch (eds.). Design and Analysis of Ecological Experiments. Oxford University Press.
57. Petratis, P.S., Beaupre S.J. and Dunham A.E. (2001). ANCOVA: Nonparametric and Randomization Approaches. pp. 116–133 in: Scheiner S.M. and Gurevitch (eds.). Design and Analysis of Ecological Experiments. Oxford University Press, Oxford.
58. Pugesek, B., Tomer A. and von Eye A. (eds.) (2002). Structural Equations Modeling: Applications in Ecological and Evolutionary Biology Research. Cambridge University Press, Cambridge, U.K.
59. Paukert, C.P. and Wittig T.A. (2002). Applications of multivariate statistical methods in fi sheries. Fisheries 27(9):16-22
60. Palmer, M.W. (1988). Fractal geometry: a tool for describing spatial patterns of plant communities. Vegetatio 75:91–102.
61. Odum, E.P. and Barrett G.W. (2005). Fundamentals of Ecology. Thomas Brooks/Cole, Belmont, CA.
62. Orloci, L. (1975). Multivariate Analysis in Vegetation Research. Junk, The Hague.
63. Rohlf, F.J. (1995). BIOM: A Package of Statistical Programs to Accompany the Text Biometry. Exeter Software, Setauket, New York.
64. Scheiner, S.M. (2001). Theories, hypotheses, and statistics. pp. 3-13 in: Scheiner S.M. and J. Gurevitch (eds.). Design and Analysis of Ecological Experiments. Oxford University Press, Oxford.
65. Shipley, B. (2000). Cause and Correlation in Biology. Cambridge University Press, Cambridge, U.K.
66. Statsoft, Inc. 1995. STATISTICA for Windows. 2nd edition. Tulsa, OK.
67. Stiling, P. (2002). Ecology: Theories and Applications. Prentice Hall, Upper Saddle River, N.J.
68. Swan, J.M.A. (1970). An examination of some ordination problems by use of simulated vegetational data. Ecology 51: 89–102.
69. Ter Braak C.J.F. & Looman C.W.N. (1986). Weighted averaging, logistic

regression and the Gaussian response model. Vegetatio, 65: 3-11
70. Ter Braak, C.J.F. (1987). The analysis of vegetation-environment relationships by canonical correspondence analysis. Vegetatio, 69:69-77.
71. Ter Braak, C. J. F. (1994). Canonical community ordination. Part I: Basic theory and linear methods. Ecoscience 1 (2), 127-140.
72. Ter Braak C.J.F. & Šmilauer P. (1998). CANOCO Reference Manual and User›s Guide to Canoco for Windows. Microcomputer Power, Ithaca, USA. 352 pp.
73. Thompson, S.K. (2002). Sampling. John Wiley & Sons, New York. Second Edition.
74. Turner MG. (1989). Landscape ecology: the effect of pattern on process. Ann. Rev. Ecol. Syst. 20:171-197.
75. Turner MG, RH Gardner and RV O'Neill (2001). Landscape Ecology in Theory and Practice: Pattern and Process. Springer, New York.
76. van der Meer, J. Heip C.H., Herman P.J.M., Moens T. and van Oevelen D. (2005). Measuring the Flow of Energy and Matter in Marine Benthic Animal Populations. pp. 326–408 in: Eleftheriou. A. and A. McIntyre (eds.). 2005. Methods for Study of Marine Benthos. Blackwell Science Ltd., Oxford, UK.
77. Van der Plogeg, S.W.F. & Vlijm L. (1978). Ecological evaluation, nature conservation and land use planning with particular reference to the methods used in the Netherlands.Biol.Consero.14:197-221.
78. Van Sickle, J. (1997). Using mean similarity dendrograms to evaluate classifications. Journal of Agricultural, Biological, and Environmental Statistics, 2:370-388.
79. Whittaker R.H. (1965). Dominance and diversity in land plant communities. Science 147: 250–260.
80. Whittaker, R.H. (1967). Gradient analysis of vegetation. Biol. Rev. 42:207-264.
81. Zare Chahouki, M. A. Azarnivand H., Jafari M. & Tavili A. (2009). Multivariate Statistical Methods as a Tool for Model_Based Prediction of Vegetation Types, Russian Journal of Ecology, 41(1): 84-94.
82. Zare Chahouki, M.A. (2006). Modeling the spatial distribution of plant species in arid and semi-arid rangelands. PhD Thesis in Range management, Faculty of Natural Resources, University of Tehran, 180 p. (In Persian).
83. Schluter. D. and Grant P.R. (1982). The distribution of Geospiza difficilis on Galapagos islands: test of three hypotheses. Evolution 36:1213-1226

Chapter 2

CONTAMINATION OF ENVIRONMENT WITH POLYCYCLIC AROMATIC HYDROCARBONS IN INDIA

Khageshwar Singh Patel[1], Shobhana Ramteke[1], Yogita Naik[1], Bharat Lal Sahu[1], Saroj Sharma[2], Jutta Lintelmann[3], Matuschek Georg[3]

[1] School of Studies in Chemistry/Environmental Science, Pt. Ravishankar Shukla University, Raipur, India

[2] Department of Chemistry, Devi Rathi Mahila Mahavidhaya, Rajnandagaon, India

[3] GSF-Forschungszentrum für Umwelt und Gesundheit, Institut für Ökologische Chemie, Neuherberg, Germany

ABSTRACT

In India is contaminated with polycyclic aromatic hydrocarbons (PAHs) due to occurring of large anthropogenic activities i.e. fuel combustion, mineral roasting and biomass burning. Hence, contamination of 13 toxic PAHs: phenanthrene, anthracene, fluoranthene, pyrene, benz (a) anthracene, benzo (b) fluoranthene, benzo (k) fluoranthene, benzo (a) pyrene, benzo (ghi) perylene, dibenz (ah) anthracene, indeno1,2,3-(cd) pyrene, coronene and coronene in the environment (i.e. ambient particulate matter, road dust, sludge and sewage) of the most industrialized area: Raipur city, India is described. The $\sum PAH_{13}$ concentration in the 16 environment materials was ranged from 7980 - 1,051,300 µg/kg with mean value of 172,613 ± 154,726 µg/kg. The concentration variations, toxicities and sources of the PAHs in various environmental compartments are discussed.

INTRODUCTION

Polycyclic aromatic hydrocarbons (PAHs) are a large group of chemical compounds, I, with a similar structure comprising two or more joined aromatic carbon rings [1]. The compounds are formed by combustion of fuels, biomass and waste materials [2]. Polycyclic aromatic compounds are carcinogenic and mutagenic compounds, causing irreversible changes in the structure and functioning of living organisms [3]. There are thousands of PAH compounds in the environment but 13 compounds i.e. naphthalene, phenanthrene, anthracene, fluoranthene, pyrene, chrysene, benz (a) anthracene, benzo (b) fluoranthene, benzo (k) fluoranthene, benzo (a) pyrene, benzo (ghi) perylene, dibenz (ah) anthracene, indeno1,2,3-(cd) pyrene and coronene of the increased environmental and health interests. These compounds differ substantially in their physical, chemical and toxicological properties and therefore, their quantification in the environment is needed. The most potent carcinogens have been shown to be benzo [a] anthracene, benzo [a] pyrene and dibenz [ah] anthracene. They are multimedia contaminants, reported at elevated levels in several environmental samples i.e. dust, particulate matter, sludge and sewage of various region of the World [4]-[26].

METHODS AND MATERIALS

Selection of Sampling Sites

Raipur (21°23'N, 81°63'E) is the capital city of the Chhattisgarh state with population of ≈2 million. The Raipur city is now becoming an important regional commercial and industrial destination for the coal, power, steel and aluminum industries. Several steel rolling mills, sponge iron plants, steel plants, agro-industries, thermal power plants and vehicles ($>1.0 \times 10^5$) are emitting effluents in and around the city.

Collection of Samples

The road dust, sludge and sewage samples were collected using a stainless-steel scoop from 13 locations of Raipur city in February 2010, Figure 1. They were kept in 250-mL glass bottle and dried at 30°C in an oven for overnight. The samples were crushed into fine particles by mortar and sieved out the particles of mesh size < 0.1 mm. The samples were stored in aluminum foil.

The coarse particulate matter (PM_{10}) and fine particulate matter ($PM_{2.5}$) were collected by using Partisol Model 2300 Sequential speciation air sampler. The sampler was installed at the roof of the building, ≈10 m above from the ground level at residential site: Dagania, Raipur. Both $PM_{2.5}$ and PM_{10} were collected simultaneously over 47 mm quartz fiber filters housed in molded filter cassette. The sampler was run for 24 hrs (6.00 am - 6.00 am) at flow rate of 10 L/min. One sample blank was used for collection of both PM_{10} and $PM_{2.5}$. The loaded filters were dismounted, brought to laboratory, and heated up to 30°C for 6 hrs to remove the moisture contents. The filters were transferred into the desiccator, and finally weighted to record the particulate contents.

Analysis of Carbons

The CHNSO-IRMS Analyzer by SV Instruments Analytica Pvt Ltd. was used for analysis of the total carbon (TC). Three carbons i.e. elemental carbon (EC), organic carbon (OC) and carbonate carbon (CC) were analyzed in the samples. The total carbn (TC) sample was oxidized with O_2 at 1020°C with constant helium flow by measuring the resulting CO_2 with thermal conductivity detector. The CC content was analyzed by treating the sample with HCl acid in the CO_2 free atmosphere. The resulting CO_2 was measured by coulometric titration method. The OC content was analyzed by titration method using $K_2Cr_2O_7$ as oxidant, and the excess of $K_2Cr_2O_7$ was determined by titration with the $FeSO_4 \cdot 7H_2O$ solution. The EC content was evaluated by using following equation.

$$EC = TC - (OC + CC)$$

Analysis of PAHs

The PAH samples were analyzed by capillary gas chromatography (Varian STAR 3400 CX) using temperature programmable splitless injection, a fused silica RTX5-MS column and ion trap mass spectrometric detection [27].

RESULTS AND DISCUSSION

Carbon Concentration

All samples are colored, ranging from brown to black, depending on the EC content. The content of EC, OC and CC in the 16 environmental shown in Table 1. Relatively high content of EC in all samples was achieved, ranging from 6.5% - 13.5% with mean value of 8.4% ± 1.1%. Very low content of OC and CC was observed in the dust, sludge and sewage samples unlikely to PM samples may be due to their degradation and water solubility, Figure 1. The EC content with the OC and CC had good relation (r = 0.94 - 0.96), indicating origin from the similar sources.

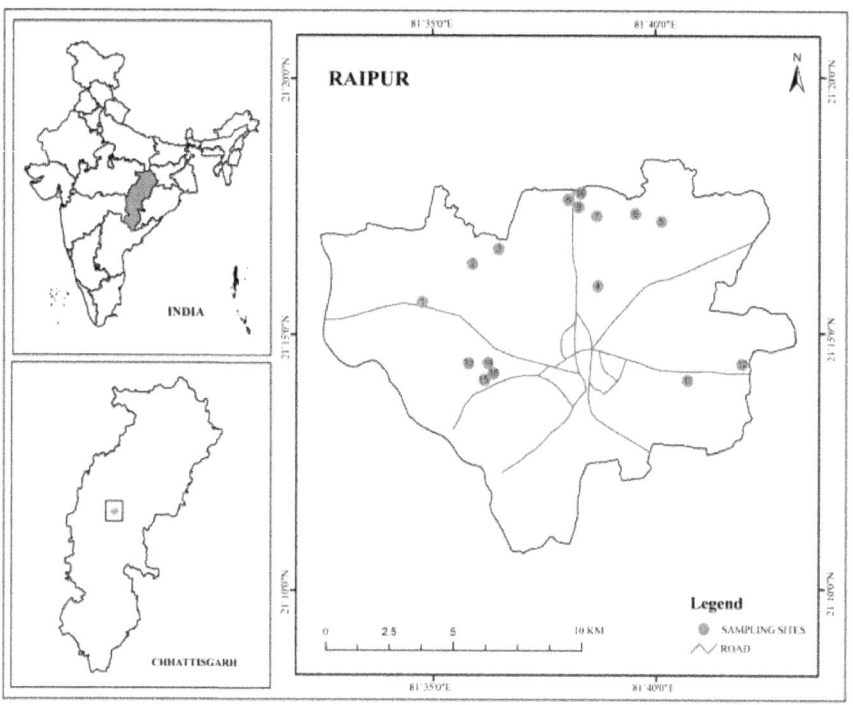

Figure 1: Representation of sampling locations.

PAHs Concentration

The chemical characteristics of 13 PAHs i.e. phenanthrene (Phe), anthracene (Ant), fluoranthene (Fla), pyrene (Pyr), benz [a] anthracene (Baa), chrysene (Cry), benzo [b] fluoranthene (Bbf), benzo [k] fluoranthene (Bkf), benzo [a] pyrene (Bap), benzo [ghi] perylene (Bgh), dibenz [a,h] anthracene (Dba), indeno [1,2,3-cd] pyrene (Ind) and coronene is summarized in Table 2. The content of 13 PAHs in 16 environmental samples is presented in Table 3. The sum of total concentration of PAHs ($\Sigma PAH13$) in the road dust of Raipur city (n = 8) was ranged from 10,427 - 26,031 µg/kg with mean value of 15,282 ± 3377 µg/kg. The highest concentration of the ΣPAH_{13} was observed at site no 5 (i.e. Birgaon) due to higher industrial and traffic emissions, Figure 2. Similarly, the concentration of ΣPAH_{13} in the SL, MW, AW and TPPW was found to be 7980, 9669, 10,570 and 8326 µg/kg, respectively. No signal for Cor was detected in the environmental samples i.e. RD, SL, MW, AW and TPPW samples. The major fraction of PAHs in the RD, AW and TPPW samples was contributed by three compounds i.e. Phe, Fla and Pyr, Figure 3. A different distribution pattern of PAHs in the SL and MW samples was observed, dominated by Pyr and Bgh contents, Figure 3. The concentration of ΣPAHs in the PM_{10} and $PM_{2.5}$ was strongly enriched, >25-folds higher than the road dust with appearing of strong Cor signal. The $PM_{2.5}$ sample was dominated by higher PAHs i.e. Bbf, Bgh and Ind, Figure 4.

Table 1: Chemical characteristics of PAHs

S. No.	PAHs	No. of rings	Abbreviation	Formula
1	Phenanthrene	3	Phe	$C_{14}H_{10}$
2	Anthracene	3	Ant	$C_{14}H_{10}$
3	Fluoranthene	4	Fla	$C_{16}H_{10}$
4	Pyrene	4	Pyr	$C_{16}H_{10}$
5	Benz [a] anthracene	4	Baa	$C_{18}H_{12}$
6	Chrysene	4	Cry	$C_{18}H_{12}$
7	Benzo [b] fluoranthene	5	Bbf	$C_{20}H_{12}$
8	Benzo [k] fluoranthene	5	Bkf	$C_{20}H_{12}$
9	Benzo [a] pyrene	5	Bap	$C_{20}H_{12}$
10	Dibenz [ah] anthracene	5	Dba	$C_{22}H_{14}$
11	Benzo [ghi] perylene	6	Bgh	$C_{22}H_{12}$
12	Indeno [1,2,3-cd] pyrene	6	Ind	$C_{22}H_{12}$
13	Coronene	6	Cor	$C_{24}H_{12}$

Table 2: Concentration of carbons and polycyclic aromatic hydrocarbons in environmental samples

S. No.	Sample type	Location	BC, %	OC, %	CC, %	ΣPAHs, mg/kg
1	RD1	Tatibandh	6.5	0.32	0.45	10.4
2	RD2	Hirapur	6.7	0.38	0.47	11.2
3	RD3	Sarora	6.9	0.34	0.41	13.1
4	RD4	Khamtarai	6.8	0.36	0.47	14.1
5	RD5	Birgaon	10.4	0.53	0.68	26.0
6	RD6	Urla	7.2	0.39	0.52	14.2
7	RD7	Sankra	8.2	0.42	0.56	16.7
8	RD8	Siltara	8.2	0.44	0.52	16.5
9	SL	Siltara	7.5	0.03	0.12	8.0
10	MW	Siltara	7.6	0.03	0.11	9.7
11	TPPW	Monate, Urla	8.9	0.04	0.17	8.3
12	AW	IGAU	9.0	0.05	0.19	10.6
13	$(PM_{10})1$	Dagania	6.8	5.8	4.6	505
14	$(PM_{10})2$	Dagania	6.7	5.6	4.2	347
15	$(PM_{2.5})1$	Dagania	13.5	8.3	5.1	1051
16	$(PM_{2.5})2$	Dagania	13.2	8.1	5.0	700

RD, SL, MW, TPPW, AW, PM_{10} and $PM_{2.5}$ represent road dust, sludge, municipal/sewage waste, thermal power plant waste, agricultural waste, coarse particulate matter and fine particulate matter, respectively.

Whereas, the PM10 sample was dominated by PAHs i.e. Fla, Pyr, Bgh and Ind, Figure 4. The PAHs content in the dust was negatively and fairly correlated with particle size (r = −0.89), Figure 5. The concentration of the PAHs in the environmental samples of studied area was found to be comparable to the other parts of the country and World [4]-[25].

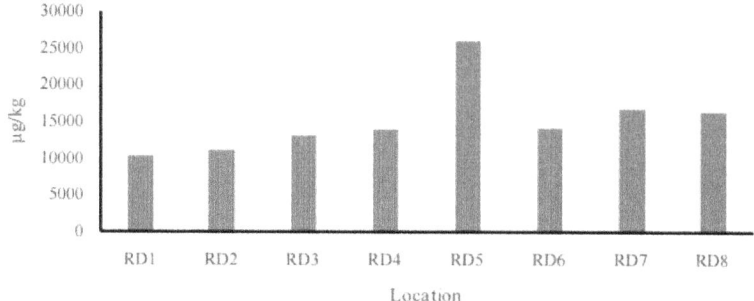

Figure 2: Spatial distribution of PAHs in the road dust.

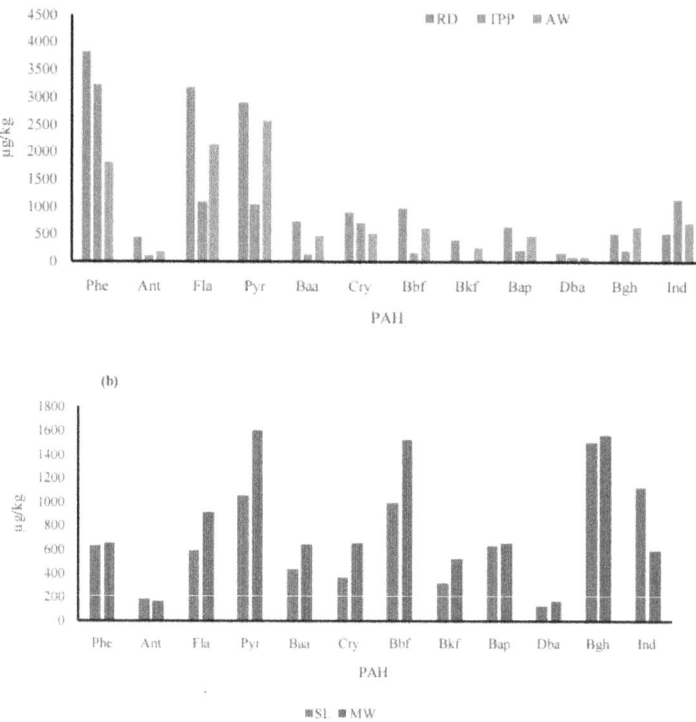

Figure 3: Distribution of PAHs in dust (RD), thermal power plant ash (TPPA), agricul-

tural waste (AW), sludge (SL) and municipal waste (MW).

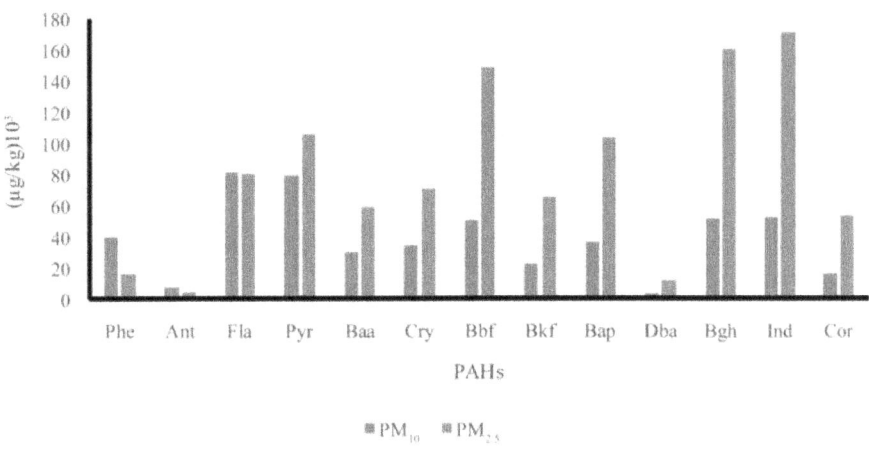

Figure 4: Distribution of PAHs in the PM_{10} and $PM_{2.5}$.

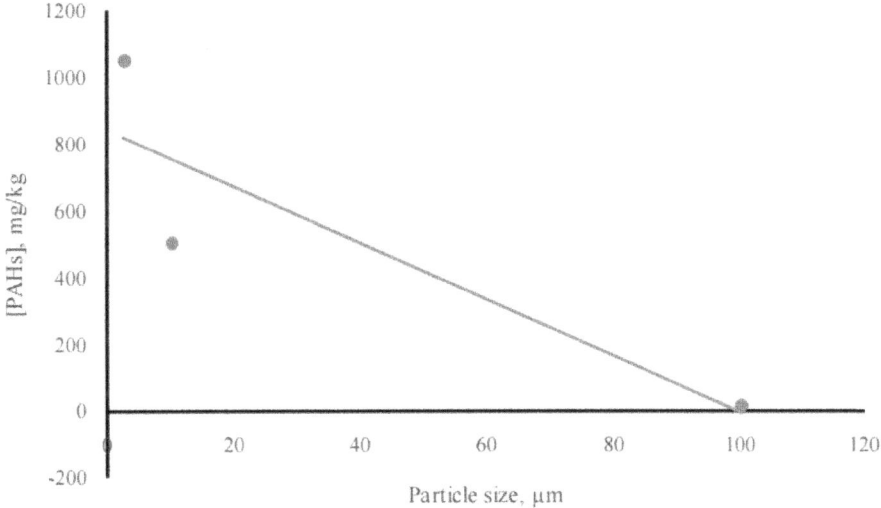

Figure 5: Correlation of PAHs content in the sample with particle size.

Table 3: Concentration of PAHs in environmental samples, μg/kg

S. No.	Phe	Ant	Fla	Pyr	Baa	Cry	Bbf	Bkf	Bap	Dba	Bgh	Ind	Cor
RD1	3168	301	1921	2222	394	622	586	227	341	18	373	254	0
RD2	2549	218	2669	2494	408	869	629	295	358	151	262	334	0
RD3	3402	320	2874	2240	600	902	866	343	439	97	602	430	0
RD4	3484	252	2208	2978	797	895	1078	442	710	28	582	632	0
RD5	5022	763	4759	5258	1886	1595	1986	880	1636	103	1019	1124	0
RD6	2636	178	3073	2254	781	1135	1358	480	732	200	670	677	0
RD7	5688	797	4655	2928	414	490	468	175	275	496	156	194	0
RD8	4622	732	3240	2974	713	826	931	398	720	226	521	550	0
SL	632	186	598	1056	439	365	990	318	637	132	1505	1122	0
MW	653	168	914	1598	642	658	1525	526	653	167	1564	601	0
TPPW	3240	120	1104	1056	154	708	168	145	204	84	204	1139	0
AW	1829	191	2149	2574	475	515	630	256	486	101	644	720	0
(PM10)1*	39.3	7.7	81.7	79.8	29.6	34.6	50.5	22.1	35.9	3.5	51.5	52.7	15.9
(PM10)2*	24.7	5.2	39.9	39.6	18.5	23.3	43.5	19.0	22.2	2.7	45.4	47.8	15.6
(PM2.5)1*	16.1	3.8	79.9	106.1	59.1	70.6	149.1	65.4	104.4	11.3	160.3	171.1	54.1
(PM2.5)2*	15.9	4.4	37.7	47.3	30.7	42.4	115.6	51.1	53.3	6.8	119.5	129.2	46.3

* = 10^3

Vertical Distribution of PAHs

The vertical distribution of the PAHs from 0 - 30 cm in the sludge samples was studied, and presented in Figure 6. The ∑PAHs content was strongly increased with increase of the sludge depth profile from 0 - 30 cm, may be due to their poor adsorption with the geo-media. Among them, extremely high vertical distribution of compounds i.e. Fla, Pyr, Bbf and Bap was observed.

Toxicities

The toxicities of PAHs increases as the mass number increases, and seven PAHs (i.e. Pyr, Baa, Bbf, Bkf, Bap, Dba and Ind) are considered to be more toxic, may bedue to higher thermal stability and delocalization of π-electrons. The carcinogenic potentiality of Pyr, Baa, Bbf, Bkf, Bpa, Dba and Ind reported was 0.01, 0.1, 0.1, 0.1, 0.1, 1.0 and 0.1, respectively [28].

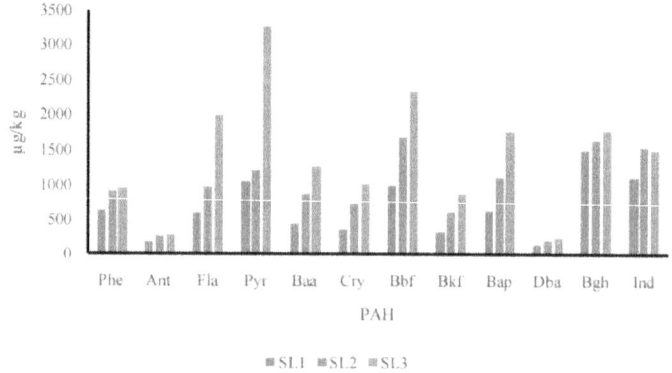

Figure 6: Vertical distribution of PAHs in the sludge, SL1 = 0 - 10 cm, SL2 = 10 - 20

cm, SL3 = 20 - 30 cm.

The benzo [a] pyrene equivalent (BapE) value was computed by using the following equation:

Total BapE $= [\sum_i C_i] \times [TEF_i]$

where, C_i and TEF_i are the concentration and the corresponding toxic equivalent factor (TEF) value of PAHs. The BapE value for RD, SL, MW, AW, TPPW, PM_{10} and $PM_{2.5}$ was found to be 1108, 1135, 1237, 846, 475, 52,000 and 138,500 µg/kg in the term of Bap. The carcinogenic fraction of PAHs in RD, SL, MW, AW, TPPW, PM_{10} and $PM2_{.5}$ samples was ranged from 5.7% - 15.7% with significantly higher value for SL, MW and PM samples, Figure 7. The concentration of PAHs in the environmental samples was found to be several folds higher than recommended value of 1000 µg/kg [29].

Correlation and Sources

The correlation matrix of the carbons and PAHs are summarized in Table 4. The PAHs had fair correlation with the BC, OC and CC contents ($r = 0.70 - 0.96$), indicating their origin from the burning processes. The lower PAHs (i.e. Phe, Ant, Fla and Pyr) among themselves had fair correlation, may be due to existence of their larger fractions in the gaseous forms, Table 4. The higher PAHs (i.e. Baa, Cry, Bbf, Bkf, Bap, Bgh and Ind except Dba) among themselves had good correlation, indicating origin from the burning processes, Table 4.

The diagnosis ratios: Phe/Antand [Fla]/[Fla + Pyr] were used to find out the sources of PAHs in the studied samples [30] [31]. The Phe/Ant ratio for TPPW, RD, AW, SL, MW, PM_{10} and $PM_{2.5}$ was found to be 27, 10.2, 9.6, 3.4, 3.9, 5.0 and 3.9, respectively, suggesting the domination of petrogenic PAHs in the TPPW, RD and AW samples, Figure 8. The Fla./Fla + Pyr. ratio of >0.5, 0.5 - >0.4 and <0.4 was used as signature for PAHs emission from combustion of grass, wood/ coal, petroleum and diesel, respectively [31]. The Fla./Fla + Pyr. ratio was ranged from 0.36 - 0.52, indicating domination of biomass or coal origin PAHs in the RD, TPPW and PM_{10} samples, Figure 8.

CONCLUSION

The light PAHs (3 - 4 ring) was found to be dominated in the RD, TPPW, AW and PM_{10} samples unlikely to SL, MW and $PM_{2.5}$ samples. Their origins were largely pyrogenic, emitted by combustion of biomass, coal and diesel. The higher PAHs (5 - 6 ring) was found to extremely enrich in the $PM_{2.5}$ sample due to origin by the combustion processes.

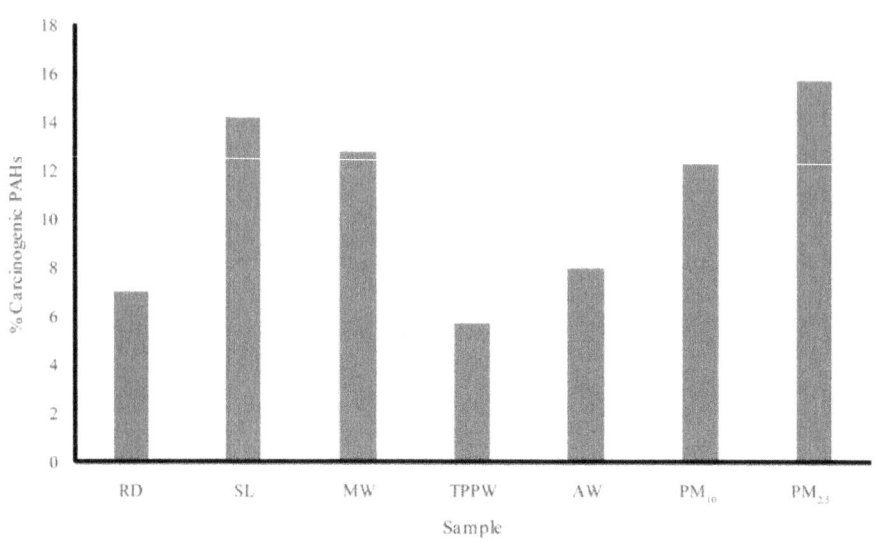

Figure 7: Percentage of carcinogenic PAHs in environmental samples.

The PHAs concentration was remarkably increased vertically and might be due to poor adsorption by the geo-media. The PAHs content in the environmental samples of the studied area was found to be several folds higher than recommended value of 1000 µg/kg.

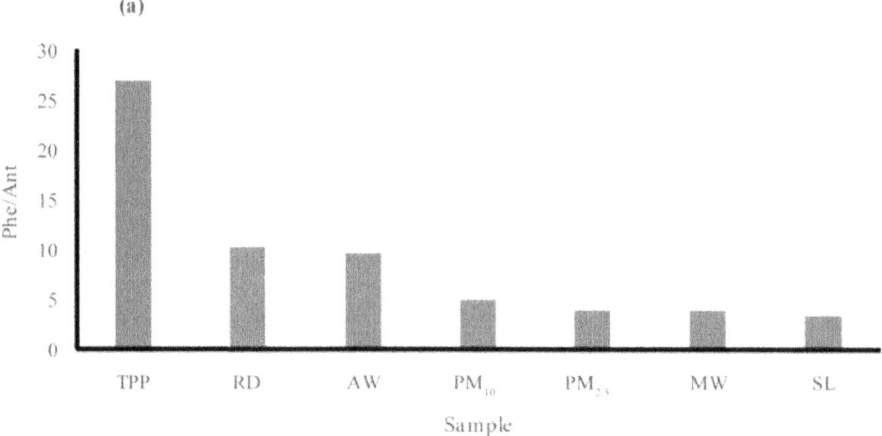

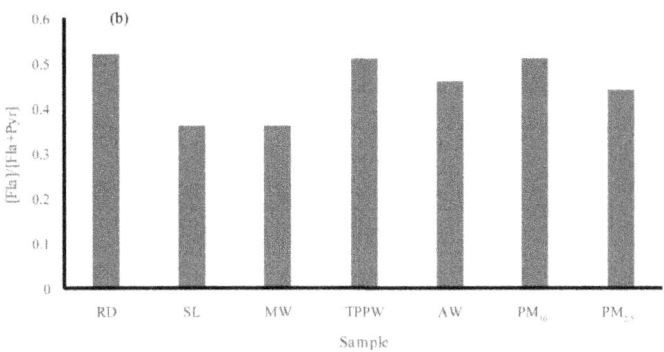

Figure 8: Diagnostic ratio for source determination of PAHs.

Table 4: Correlation matrix of PAHs in the road dust

	Phe	Ant	Fla	Pyr	Baa	Cry	Bbf	Bkf	Bap	Dba	Bgh	Ind
Phe	1											
Ant	0.96	1										
Fla	0.80	0.81	1									
Pyr	0.61	0.64	0.69	1								
Baa	0.35	0.39	0.54	0.91	1							
Cry	-0.02	0.06	0.34	0.71	0.91	1						
Bbf	0.12	0.17	0.39	0.75	0.95	0.96	1					
Bkf	0.18	0.24	0.42	0.83	0.98	0.97	0.98	1				
Bap	0.31	0.37	0.48	0.89	0.99	0.92	0.96	0.99	1			
Dba	0.60	0.57	0.66	-0.01	-0.23	-0.37	-0.31	-0.33	-0.26	1		
Bgh	0.05	0.10	0.24	0.65	0.90	0.92	0.96	0.94	0.91	-0.47	1	
Ind	0.15	0.20	0.37	0.78	0.96	0.95	0.99	0.99	0.98	-0.33	0.95	1

ACKNOWLEDGEMENTS

We are thankful the CCOST, Raipur for granting research support for doing this work.

REFERENCES

1. Bjorseth, A. and Dennis, A. (1980) Proceedings of International Polynuclear Aromatic Hydrocarbon Symposia, PAHIV: Polynuclear Aromatic Hydrocarbons, Chemistry and Biological Effects. Battelle Press, Columbus, OH, 1125 p.

2. Finlayson-Pitts, B.J. and Pitts, J.N. (2000) Chemistry of the Upper and Lower Atmosphere. Academic Press, San Diego.

3. Pickering, R.W. (1999) A Toxicological Review of Polycyclic Aromatic Hydrocarbons. Journal of Toxicology—Cutaneous and Ocular

Toxicology, 18, 101-135. http://dx.doi.org/10.3109/15569529909037562
4. Baran, S. and Oleszczuk, P. (2003) The Concentration of Polycyclic Aromatic Hydrocarbons in Sewage Sludge in Relation to the Amount and Origin of Purified Sewage. Polish Journal of Environmental Studies, 12, 523-529.
5. Pérez, S., Farré, M., Jesús Garcí a, M. and Barceló, D. (2001) Occurrence of Polycyclic Aromatic Hydrocarbons in Sewage Sludge and their Contribution to its Toxicity in the ToxAlert® 100 Bioassay. Chemosphere, 45, 705-712. http://dx.doi.org/10.1016/S0045-6535(01)00152-7
6. Blanchard, M., Teil, M.J., Ollivon, D., Legenti, L. and Chevreuil, M. (2004) Polycyclic Aromatic Hydrocarbons and Polychlorobiphenyls in Wastewaters and Sewage Sludges from the Paris Area (France). Environmental Research, 95, 184-197. http://dx.doi.org/10.1016/j.envres.2003.07.003
7. Harrison, E.Z., Oakes, S.R., Hysell, M. and Hay, A. (2006) Organic Chemicals in Sewage Sludge. Science of Total Environment, 367, 481-497. http://dx.doi.org/10.1016/j.scitotenv.2006.04.002
8. Cai, Q.Y., Mo, C.H., Wu, Q.T., Zeng, Q.Y. and Katsoyiannis, A. (2007) Occurrence of Organic Contaminants in Sewage Sludges from Eleven Wastewater Treatment Plants, China. Chemosphere, 68, 1751-1762. http://dx.doi.org/10.1016/j.chemosphere.2007.03.041
9. Lorenzi, D., Entwistle, J.A., Cave, M. and Dean, J.R. (2011) Determination of Polycyclic Aromatic Hydrocarbons in Urban Street Dust: Implications for Human Health. Chemosphere, 83, 970-977. http://dx.doi.org/10.1016/j.chemosphere.2011.02.020
10. Tang, X., Shen, C., Cheema, S.A., Chen, L., Xiao, X., Zhang, C., Liu, W., Li, F. and Chen, Y. (2010) Levels and Distributions of Polycyclic Aromatic Hydrocarbons in Agricultural Soils in an Emerging e-waste Recycling town in Taizhou Area, China. Journal of Environmental Science and Health Part A, Toxic/Hazardous Substances and Environmental Engineering, 45, 1076-1084. http://dx.doi.org/10.1080/10934529.2010.486336
11. Amuda, O.S. and Adelowo-Imeokparia, F.E. (2007) Polycyclic Aromatic Hydrocarbons in Municipal Waste Ashes from Three Waste Dumps in Lagos, Nigeria. Bulletin of the Chemical Society of Ethiopia, 21, 141-144. http://dx.doi.org/10.4314/bcse.v21i1.61402
12. Pengchai, P., Nakajima, F. and Furumai, H. (2005) Estimation of Origins of Polycyclic Aromatic Hydrocarbons in Size-Fractionated Road Dust in Tokyo with Multivariate Analysis. Water Science and Technology, 51,

169-175.

13. Boonyatumanond, R., Murakami, M., Wattayakorn, G., Togo, A. and Takada, H. (2007) Sources of Polycyclic Aromatic Hydrocarbons (PAHs) in Street Dust in a Tropical Asian Mega-City, Bangkok, Thailand. Science of the Total Environment, 384, 420-432. http://dx.doi.org/10.1016/j.scitotenv.2007.06.046

14. Naspinski, C., Lingenfelter, R., Cizmas, L., Naufal, Z., He, L.Y., Islamzadeh, A., Li, Z.W., Li, Z., McDonald, T. and Donnelly, K.C. (2008) A Comparison of Concentrations of Polycyclic Aromatic Compounds Detected in Dust Samples from Various Regions of the World. Environment International, 34, 988-993. http://dx.doi.org/10.1016/j.envint.2008.03.008

15. Hassanien, M.A. and Abdel-Latif, N.M. (2008) Polycyclic Aromatic Hydrocarbons in Road Dust over Greater Cairo, Egypt. Journal of Hazardous Materials, 151, 247-254. http://dx.doi.org/10.1016/j.jhazmat.2007.05.079

16. Han, B., Bai, Z.P., Guo, G.H., Wang, F., Li, F., Liu, Q.X., Ji, Y.Q., Li, X. and Hu, Y.D. (2009) Characterization of PM10 Fraction of Road Dust for Polycyclic Aromatic Hydrocarbons (PAHs) from Anshan, China. Journal of Hazardous Materials, 170, 934-940. http://dx.doi.org/10.1016/j.jhazmat.2009.05.059

17. Wang, W., Huang, M.J., Kang, Y., Wang, H.S., Leung, A.O., Cheung, K.C. and Wong, M.H. (2011) Polycyclic Aromatic Hydrocarbons (PAHs) in Urban Surface Dust of Guangzhou, China: Status, Sources and Human Health Risk Assessment. Science of the Total Environment, 409, 4519-4527. http://dx.doi.org/10.1016/j.scitotenv.2011.07.030

18. Yu, B., Xie, X., Ma, L.Q., Kan, H. and Zhou, Q. (2013) Source, Distribution, and Health Risk Assessment of Polycyclic Aromatic Hydrocarbons in Urban Street Dust from Tianjin, China. Environmental Science and Pollution Research, 21, 2817-2825. http://dx.doi.org/10.1007/s11356-013-2190-z

19. Mohd Radzi, N.A.S., Bakar, N.K.A., Emenike, C.U. and Abas, M.R. (2015) Polycyclic Aromatic Hydrocarbons (PAHs): Contamination Level and Risk Assessment in Urban Areas, Kuala Lumpur, Malaysia. Desalination and Water Treatment. http://dx.doi.org/10.1080/19443994.2015.1021103

20. Martuzevicius, D., Kliucininkas, L., Prasauskas, T., Krugly, E., Kauneliene, V. and Strandberg, B. (2011) Resuspension of Particulate Matter and PAHs from Street Dust. Atmospheric Environment, 45, 310-

317. http://dx.doi.org/10.1016/j.atmosenv.2010.10.026

21. Sharma, H., Jain, V.K. and Khan, Z.H. (2007) Characterization and Source Identification of Polycyclic Aromatic Hydrocarbons (PAHs) in the Urban Environment of Delhi. Chemosphere, 66, 302-310. http://dx.doi.org/10.1016/j.chemosphere.2006.05.003

22. Hassan, Y. and Sarma, H.P. (2013) Distribution of Polycyclic Aromatic Hydrocarbons (PAHS) in Roadside Soils from Industrial and High Traffic Area of Greater Guwahati City, Assam, India. Archives of Applied Science Research, 5, 85-93.

23. Ray, S., Khillare, P.S., Kim, K.H. and Brown, R.J. (2012) Distribution, Sources, and Association of Polycyclic Aromatic Hydrocarbons, Black Carbon, and Total Organic Carbon in Size-Segregated Soil Samples Along a BackgroundUrban-Rural Transect. Environmental Engineering Science, 29, 1008-1019. http://dx.doi.org/10.1089/ees.2011.0323

24. Cheng, H.R., Deng, Z.M., Chakraborty, P., Liu, D., Zhang, R.J., Xu, Y., Luo, C.L., Zhang, G. and Li, J. (2013) A Comparison Study of Atmospheric Polycyclic Aromatic Hydrocarbons in Three Indian Cities Using PUF Disk Passive Air Samplers. Atmospheric Environment, 73, 16-21. http://dx.doi.org/10.1016/j.atmosenv.2013.03.001

25. Fu, F.F., Tian, B.Y., Lin, G.S., Chen, Y.Q. and Zhang, J.H. (2010) Chemical Characterization and Source Identification of Polycyclic Aromatic Hydrocarbons in Aerosols Originating from Different Sources. Journal of the Air and Waste Management Association, 60, 1309-1314. http://dx.doi.org/10.3155/1047-3289.60.11.1309

26. Wingfors, H., Hägglund, L. and Magnusson, R. (2011) Characterization of the Size-Distribution of Aerosols and Particle-Bound Content of Oxygenated PAHs, PAHs, and n-Alkanes in Urban Environments in Afghanistan. Atmospheric Environment, 45, 4360-4369. http://dx.doi.org/10.1016/j.atmosenv.2011.05.049

27. Lintelmann, J., Fischer, K., Karg, E. and Schroppel, A. (2005) Determination of Selected Polycyclic Aromatic Hydrocarbons and Oxygenated Polycyclic Aromatic Hydrocarbons in Aerosol Samples by High-Performance Liquid Chromatography and Liquid Chromatography-Tandem Mass Spectrometry. Analytical and Bioanalytical Chemistry, 381, 508-519. http://dx.doi.org/10.1007/s00216-004-2883-8

28. EPA (2005) Air Toxics Hot Spots Program Risk Assessment Guidelines, Part II Technical Support Document for Describing Available Cancer Potency Factors. Office of Environmental Health Hazard Assessment, California Environmental Protection Agency, Sacramento.

29. CCME (Canadian Council of Ministers of the Environment) (2008) Canadian Soil Quality Guidelines for Carcinogenic and Other Polycyclic Aromatic Hydrocarbons (PAHs). Environmental and Human Health Effects, Scientific Supporting Document, 218 p.
30. Magi, E., Bianco, R., Ianni, C. and Carro, M.D. (2000) Distribution of Polycyclic Aromatic Hydrocarbons in the Sedi-ments of the Adriatic Sea. Environmental Pollution, 119, 91-98. http://dx.doi.org/10.1016/S0269-7491(01)00321-9
31. Li, G.C., Xia, X.H., Yang, Z.F., Wang, R. and Voulvoulis, N. (2006) Distribution and Sources of Polycyclic Aromatic Hydrocarbons in the Middle and Lower Reaches of the Yellow River, China. Environmental Pollution, 144, 985-993. http://dx.doi.org/10.1016/j.envpol.2006.01.047

Chapter 3

ENVIRONMENT AND HEALTH IN CONTAMINATED SITES: THE CASE OF TARANTO, ITALY

Roberta Pirastu,[1] Pietro Comba,[2] Ivano Iavarone,[2] Amerigo Zona,[2] Susanna Conti,[3] Giada Minelli,[3] Valerio Manno,[3] Antonia Mincuzzi,[4] Sante Minerba,[4] Francesco Forastiere,[5] Francesca Mataloni,[5] and Annibale Biggeri[6,7]

[1]Department of Biology and Biotechnologies Charles Darwin, Sapienza Rome University, Piazzale Aldo Moro 5, 00185 Rome, Italy

[2]Department of Environment and Primary Prevention, National Health Institute, Viale Regina Elena 299, 00161 Rome, Italy

[3]Unit of Statistics of the National Health Institute, National Center for Epidemiology, Surveillance and Health Promotion, Viale Regina Elena 299, 00161 Rome, Italy

[4]Taranto Local Health Unit, Epidemiological and Statistical Unit, Viale Virgilio 31, 74121 Taranto, Italy

[5]Department of Epidemiology, Lazio Regional Health Service, Via di Santa Costanza 53, 00198 Rome, Italy

[6]Biostatistics Unit, ISPO Cancer Research and Prevention Institute, Via Cosimo il Vecchio 2, 50139 Florence, Italy

[7]Annibale Biggeri Department of Statistics "G. Parenti", University of Florence, Viale Morgagni 59, 50134 Firenze, Italy

ABSTRACT

The National Environmental Remediation programme in Italy includes sites with documented contamination and associated potential health impacts (National Priority Contaminated Sites—NPCSs). SENTIERI Project, an extensive investigation of mortality in 44 NPCSs, considered the area of Taranto, a NPCS where a number of polluting sources are present. Health indicators available at municipality level were analyzed, that is, mortality (2003–2009), mortality time trend (1980–2008), and cancer incidence (2006-2007). In addition, the cohort of individuals living in the area was followed up to evaluate mortality (1998–2008) and morbidity (1998–2010) by district of residence. The results of the study consistently showed excess risks for a number of causes of death in both genders, among them: all causes, all cancers, lung cancer, and cardiovascular and respiratory diseases, both acute and chronic. An increased infant mortality was also observed from the time trends analysis. Mortality/morbidity excesses were detected in residents living in districts near the industrial area, for several disorders including cancer, cardiovascular, and respiratory diseases. These coherent findings from different epidemiological approaches corroborate the need to promptly proceed with environmental cleanup interventions. Most diseases showing an increase in Taranto NPCS have a multifactorial etiology, and preventive measures of proven efficacy (e.g., smoking cessation and cardiovascular risk reduction programs, breast cancer screening) should be planned. The study results and public health actions are to be communicated objectively and transparently so that a climate of confidence and trust between citizens and public institutions is maintained.

INTRODUCTION

Contaminated sites are extensively present in Europe where approximately 250,000 sites require cleanup interventions, as listed by the European Environment Agency [1]. Several thousands of these sites are located in Italy, and a total of 57 sites, defined in 2009 as National Priority Contaminated Sites (NPCSs), qualify for remediation because of contamination documented in qualitative and/or quantitative terms, and because of a potential health impact.

The site of Taranto, located in Apulia region (southern Italy), includes two municipalities and 216,618 inhabitants at 2001 Census. This site is of interest because of several polluting sources, such as a large steel plant, a refinery, the harbor, and both controlled and illegal waste dumps.

Previous environmental and epidemiological investigations in the area have provided evidence of environmental contamination [2–9]. These studies have documented severe air pollution originating mainly from the steel industry, that is, particulate matter, heavy metals, polycyclic aromatic hydrocarbons, and organ-halogenated compounds.

Epidemiological studies showed an increased mortality/morbidity from respiratory, cardiovascular diseases and several cancer sites [10–13].

SENTIERI Project (epidemiological study of residents in National Priority Contaminated Sites—NPCSs) funded by the Italian Ministry of Health studied mortality for 63 causes among residents of 44 NPCSs included in the "National Environmental Remediation Programme" [14].

The distinguishing feature of SENTIERI Project is that the epidemiological evidence evaluation has been carried out before the study to minimize the risk for researchers to be data-driven when performing and interpreting the specific epidemiologic investigation. SENTIERI dealt with the complexity of the relation between area contamination and health effects by examining, for each combination of causes of death/environmental exposures, the epidemiological evidence (1998–2009) and then building a matrix of the a priori evaluation of the strength of the causal association. The environmental exposures were classified as chemicals, petrochemicals and refineries, steel plants, power plants, mines and/or quarries, harbour areas, asbestos or other mineral fibers, landfills, and incinerators—labelled on the basis of the legislative decrees defining the sites' boundaries. A standardized procedure was set up to collect the available epidemiological literature, which was reviewed on the basis of explicit criteria and led to classify each cause of death/environmental exposures combination in terms of strength of causal association. The evaluation was categorized as sufficient to infer the presence of a causal association (S), limited to infer the presence of a causal association (L), and inadequate to infer the presence or the absence of a causal association (I). The rationale, scope, methods, and details of the a priori evidence evaluation can be found in [15], and the procedures and results of the evidence evaluation have been published in Italian [16].

With specific reference to the environmental exposures in Taranto NPCS, this procedure led to classify the presence of a steel industry, a refinery, a harbour area, and a number of landfills and waste dump sites as associated, with limited evidence, with an increased risk of lung cancer, pleural mesothelioma, nonmalignant respiratory diseases, congenital malformations, and perinatal conditions. In this context, residence near steel industry was evaluated as specifically associated with the occurrence of both acute and chronic respiratory diseases in adults and children, based on studies by [17–23]. It should be underlined, however, that air pollution from particulate matter has been causally linked by WHO with several health effects, including all-cause mortality and cardiovascular and respiratory morbidity [24].

SENTIERI analyzed mortality at municipality level in the period 1995–2002, computing standardized mortality ratios (SMR) both crude and adjusted for a deprivation index [25]. While SENTIERI strengths are the a prioriepidemiological evidence evaluation and the mortality analysis of all NPCSs adopting the same analytical approach and adjusting for deprivation, there are several limitations that should be noted, such as its ecological design and the use of mortality data at municipal level for a short period of time.

With the aim of overcoming the above limitations, this paper presents an epidemiological profile of Taranto NPCS residents analyzing different health indicators available at municipality level, that is, cause-specific mortality (2003–2009), mortality time trend (1980–2008), and cancer incidence (2006-

2007). A cohort study of the resident population examined mortality (1998–2008) and morbidity (1998–2010) in the districts close to the steel plant.

MATERIAL AND METHODS

Data Source

Details on the codes used during the study period for the 9th and 10th revisions of the International Classification of Diseases (ICD-9 and ICD-10) and on the demographic data of the two municipalities included in Taranto NPCS are presented in Appendix A.

Mortality 2003–2009

Mortality in Taranto NPCS residents was initially studied for the period 1995–2002 [12] and then updated for the years 2003–2009 (note that the period 2004-2005 was not available from ISTAT). The analysis considered 63 single or grouped causes (all ages, both genders); 0-1 and 0–14 age classes were also analyzed for a selection of causes (both genders combined). Standardized mortality ratios both crude (SMRs) and adjusted for deprivation together with 90% confidence intervals (90% CIs) were computed using regional rates for comparison [14]. In SENTIERI Project the deprivation index (DI) was constructed using the 2001 national census variables representing the following socioeconomic domains: education, unemployment, dwelling ownership, and overcrowding. The strengths and weaknesses of SENTIERI ID, its correlation with 2001 national deprivation index, its efficacy in representing deprivation in different categories of demographic dimensions, together with suggestions about the use of socioeconomic indices in small area studies of environment and health are discussed in Pasetto et al. 2011 [26].

Mortality Time Trend 1980–2008

Mortality was analyzed for a twenty-seven-year period: 1980–2008. The analysis was performed for the population 0–99 years separately for men and women; directly standardized death rates (SDRs) per 100,000 were calculated (standard: Italian population at 2001 Census) together with their 90% CI. The population size of Taranto NPCS is relatively small; therefore the study period was divided into three-year periods to obtain stable values of the indicators. The overall mortality was analyzed together with specific diseases, selected on the basis of the a priori evidence evaluation of their link with environmental exposures in Taranto NPCS. The selected diseases were all cancers (in particular lung cancer), circulatory diseases (in particular ischemic heart disease), and respiratory diseases (in particular, the acute, and chronic ones). We also analyzed the overall infant mortality (i.e. mortality from all causes during the first year of life) without gender distinction, as not informative in this age group.

Cancer Incidence 2006-2007

For cancer incidence (2006-2007), standardized incidence ratio (SIR) and 90% CI were calculated for both genders; the incidence rates of Italian South and Islands Cancer Registries macroarea (2005–2007) and of Taranto Province, excluding NPCSs municipalities (2006-2007), were used for comparison.

Mortality (1998–2008) and Morbidity (1998–2010) of the Residential Cohort

A cohort study design was applied to evaluate cause-specific mortality and hospitalization in relation to residence in specific districts close to the industrial sites. A cohort of residents (all subjects living in Taranto, Massafra, and Statte as January 1, 1998, and subsequently entered in these municipalities up to 2010) was enrolled from the municipal register. Individual follow-up for vital status assessment at 31.01.2010 was performed using municipality data. This cohort population is different from the base population analyzed for mortality in 2003–2009 (see previous paragraph for details). The socioeconomic position level (SEP) of the census block of residence and the district of residence were assigned to each participant (five categories from low to high SEP), on the basis of the addresses geocoded at the beginning of the follow-up. Occupational history for all cohort members was traced through the national insurance company (INPS) database (people employed in 1974 and subsequently), and the subcohort of individuals employed in industries located in the area was identified. Mortality/morbidity information was retrieved from Regional Health Databases (1998–2008 for mortality, 1998–2010 for hospital admissions). The associations of district of residence with mortality/morbidity were estimated by calculating mortality and morbidity hazard ratios (HR, CI 95%) using the proportional Cox models. All models considered age (temporal axis), calendar period, and area-based socioeconomic status [25], and they were calculated separately for men and women [13].

RESULTS

Mortality 2003–2009-SENTIERI

Tables 1–3 show mortality results in the periods 1995–2002 and 2003–2009.

Table 1: SENTIERI—Taranto NPCS. Mortality for the main causes of death. Number of observed cases (Obs), standardized mortality ratio crude (SMR), and adjusted for deprivation (SMR ID); 90% IC: 90% confidence interval; regional references: 1995–2002 and 2003–2009. Males and females

	1995-2002						2003-2009*					
	Males			Females			Males			Females		
Causes of death (ICD IX)	OBS	SMR (90% CI)	SMR ID (90% CI)	OBS	SMR (90% CI)	SMR ID (90% CI)	OBS	SMR (90% CI)	SMR ID (90% CI)	OBS	SMR (90% CI)	SMR ID (90% CI)
All causes (1-999.9)	7585	109 (107-111)	107 (105-109)	7104	107 (105-109)	107 (105-109)	4936	114 (111-117)	111 (108-113)	4847	108 (105-110)	107 (104-109)
All neoplasms (140.0-239.9)	2529	115 (112-119)	113 (109-116)	1716	113 (108-117)	112 (108-117)	1650	114 (110-119)	111 (106-115)	1208	113 (108-118)	111 (106-116)
Diseases of the circulatory system (401.0-405.9)	2654	105 (102-108)	103 (99-106)	3118	101 (98-104)	100 (97-103)	1645	114 (109-119)	109 (105-114)	1968	104 (100-108)	103 (99-107)
Diseases of the respiratory system (460.0-519.9)	666	107 (100-114)	107 (100-114)	406	113 (104-123)	111 (102-120)	447	117 (108-126)	112 (103-121)	268	104 (94-115)	105 (95-116)
Diseases of the digestive system (520.0-579.9)	442	114 (105-123)	114 (106-124)	472	142 (132-153)	141 (131-153)	283	147 (133-162)	136 (123-150)	233	119 (106-132)	117 (104-130)
Diseases of the genitourinary system (580.0-629.9)	101	92 (78-109)	97 (82-115)	107	89 (75-104)	91 (77-108)	71	94 (77-115)	101 (82-123)	85	89 (74-107)	87 (72-104)

*2004-2005 not available from ISTAT.

Table 2: SENTIERI—Taranto NPCS. Mortality for causes of death with limited evidence of association with environmental exposures in Taranto NPCS. Number of observed cases (Obs), standardized mortality ratio crude (SMR), and adjusted for deprivation (SMR ID); 90% IC: 90% confidence interval; regional references: 1995–2002 and 2003–2009. Males and females

	1995-2002						2003-2009*					
	Males			Females			Males			Females		
Causes of death (ICD IX)	OBS	SMR (90% CI)	SMR ID (90% CI)	OBS	SMR (90% CI)	SMR ID (90% CI)	OBS	SMR (90% CI)	SMR ID (90% CI)	OBS	SMR (90% CI)	SMR ID (90% CI)
Malignant neoplasm of trachea bronchus and lung (162.0-162.9)	840	130 (122-137)	119 (112-126)	121	135 (115-157)	130 (111-151)	516	133 (124-143)	123 (114-132)	97	130 (109-153)	121 (101-143)
Malignant pleural neoplasm (163.0-163.9)	83	521 (430-625)	293 (242-352)	14	242 (147-379)	190 (115-297)	44	519 (397-667)	272 (208-350)	12	311 (180-505)	210 (121-340)
Diseases of the respiratory system (460.0-519.9)	666	107 (100-114)	107 (100-114)	406	113 (104-123)	111 (102-120)	447	117 (108-126)	112 (103-121)	268	104 (94-115)	105 (95-116)
Acute diseases of the respiratory system (460.0-466.9, 480.0-487.9)	125	156 (134-181)	149 (127-173)	135	145 (125-167)	138 (119-159)	50	136 (106-172)	137 (107-174)	58	112 (89-140)	115 (91-143)
Chronic diseases of the respiratory system (491.0-492.9, 494.0-496.9)	388	96 (88-105)	97 (89-105)	151	92 (80-105)	92 (80-105)	322	116 (106-127)	110 (100-121)	149	104 (90-119)	100 (87-114)
Asthma (493.0-493.9)	9	41 (22-72)	42 (22-73)	11	73 (41-121)	68 (38-113)	0			1	25 (1-118)	29 (1-137)

*2004-2005 not available from ISTAT.

Table 3: SENTIERI—Taranto NPCS. Mortality for causes of death with limited evidence of association with environmental exposures in Taranto NPCS. Number of observed cases (Obs), standardized mortality ratio crude (SMR), and adjusted for deprivation (SMR ID); 90% IC: 90% confidence interval; regional references: 1995–2002 and 2003–2009. Males and females combined

Causes of death (age classes) (ICD IX)	OBS	1995-2002 Total		OBS	2003-2009* Total	
		SMR (90% CI)	SMR ID (90% CI)		SMR (90% CI)	SMR ID (90% CI)
Congenital anomalies (all ages) (740.0-759.9)	59	115 (91-142)	117 (93-145)	20	82 (54-119)	93 (62-135)
Certain conditions originating in the perinatal period (0-1) (760.0-779.9)	79	135 (111-162)	121 (100-146)	37	165 (123-218)	147 (110-193)
Acute diseases of the respiratory system (0-14) (460.0-466.9, 480.0-487.9)	4	96 (33-219)	95 (33-219)	<3	—	—
Asthma (0-14) (493.0-493.9)	<3	—	—	<3	—	—

*2004-2005 not available from ISTAT.

In Table 1, mortality, from the main causes of death, is displayed for descriptive purposes; Tables 2 and 3 present the results for the causes selected on the basis of the a priori evidence evaluation, the distinguishing feature of SENTIERI Project [15].

Table 1 shows that in both periods, for both genders, for all causes and all neoplasms, there was an excess of mortality ranging between 7% and 15%; adjustment for deprivation did not substantially change SMRs values. In both periods, among males and females, the observed mortality was above expected for circulatory, respiratory, and digestive systems diseases; also in these cases, accounting for socioeconomic factors did not essentially change the study results. For diseases of the genitourinary system, the observed mortality was similar to the expected one.

Table 2 presents the results for the causes of death for which SENTIERI classified the epidemiological evidence of causal association with the environmental exposures in Taranto NPCS as "Limited". From now on, reference will only be made to results adjusted for deprivation. Among males, lung cancer showed a 20% excess in the first period, confirmed in the second one; among females the excesses were, respectively, about 30% (1995–2002) and 20%. Correspondingly, in the two periods excesses for pleural tumors were 193% and 167% among males, 90% and 103% among females. Excesses for acute respiratory diseases among males were 49% (1995–2002) and 37% (2003–2009), and for females 38% and 14%, respectively. The observed mortality for chronic respiratory diseases in 1995–2002 was as expected in both genders, while in 2003–2009 a 10% excess was present for males. Asthma mortality was not increased, but the observed number of death is small.

Table 3 displays the results combined for males and females, again for causes with limited epidemiological evidence of causal association with the environmental exposures in Taranto NPCS. The mortality from congenital anomalies showed a 17% excess in 1995–2002, while in 2003–2009 it was below expectation. For mortality from perinatal conditions, an heterogeneous group of diseases affecting fetus or newborn spanning from pregnancy and delivery complications to digestive or hematological disorders, the excess was 21% and 47%, in the first and second periods, respectively. In the age class 0–14 less than 3 deaths from acute respiratory diseases and asthma were observed.

Some noteworthy results should be considered (results not shown in Tables). Among males, the observed deaths were above expected in 1995–2002 and 2003–2009 for dementia (resp., 105 and 102 deaths), hypertensive diseases (resp., 307 and 287 deaths), ischemic heart diseases (resp., 1032 and 679 deaths), and cirrhosis (resp., 266 and 156 deaths). In 2003–2009 excesses were reported for melanoma (50%, 26 deaths), non-Hodgkin lymphoma (34%, 45 deaths), and myeloid leukemia (35%, 37 deaths).

Mortality Time Trend Analysis 1980–2008

Time trends for mortality from all causes and selected causes among adults and overall infant mortality are presented in Figures 1–17 and in Appendix B (see Tables S2, S3, and S4 in Supplementary Material available online at http://dx.doi.org/10.1155/2013/753719).

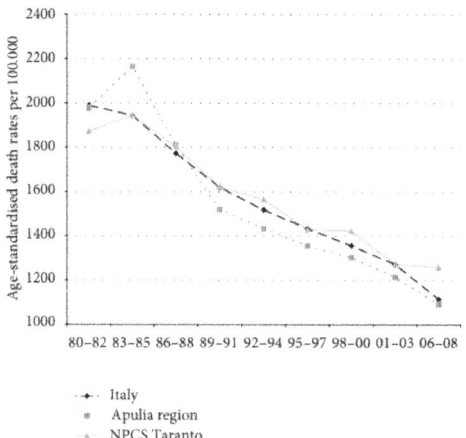

Figure 1: Overall mortality. Trends in age-standardised (Italian Census 2001) death rates per 100.000 (1980–2008) (2004-2005 data were not available). All ages. Men.

Environment and Health in Contaminated Sites: The Case of Taranto, Italy   59

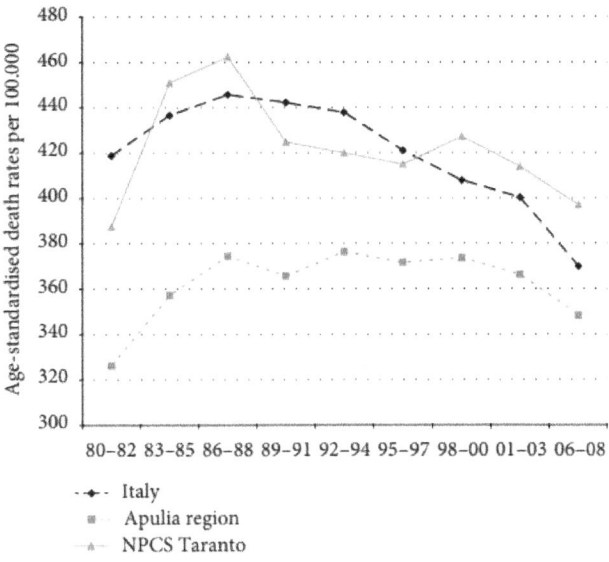

Figure 2: All cancers. Trends in age-standardised (Italian Census 2001) death rates per 100.000, from selected causes of death (1980–2008) (2004-2005 data were not available). All ages. Men.

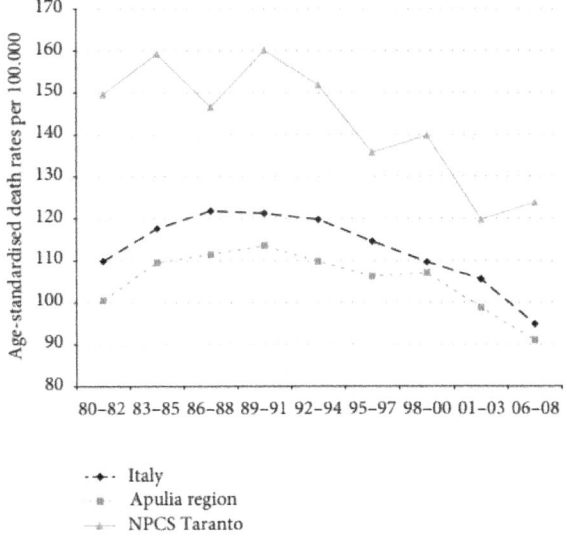

Figure 3: Lung cancer. Trends in age-standardised (Italian Census 2001) death rates per 100.000, from selected causes of death (1980–2008) (2004-2005 data were not available). All ages. Men.

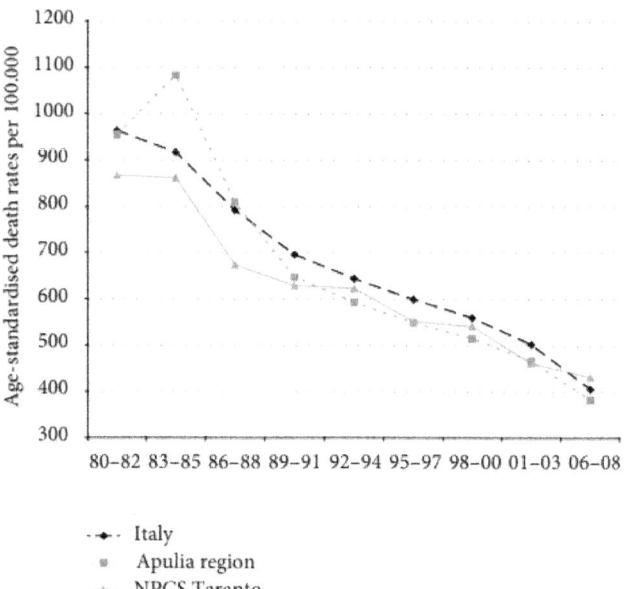

Figure 4: Circulatory diseases. Trends in age-standardised (Italian Census 2001) death rates per 100.000, from selected causes of death (1980–2008) (2004-2005 data were not available). All ages. Men.

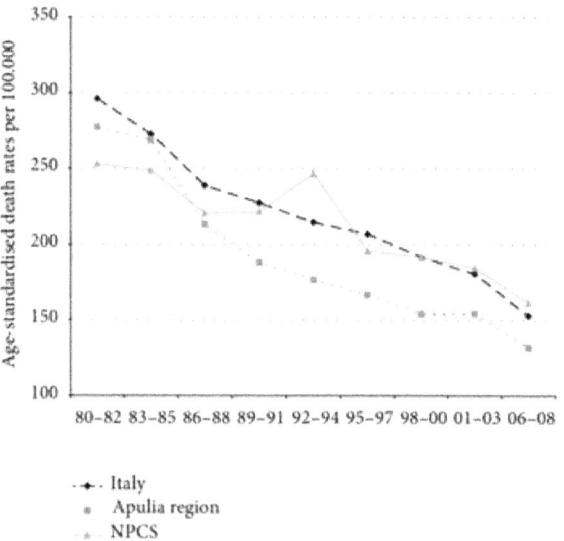

Figure 5: Ischaemic heart diseases. Trends in age-standardised (Italian Census 2001) death rates per 100.000, from selected causes of death (1980–2008) (2004-2005 data were not available). All ages. Men.

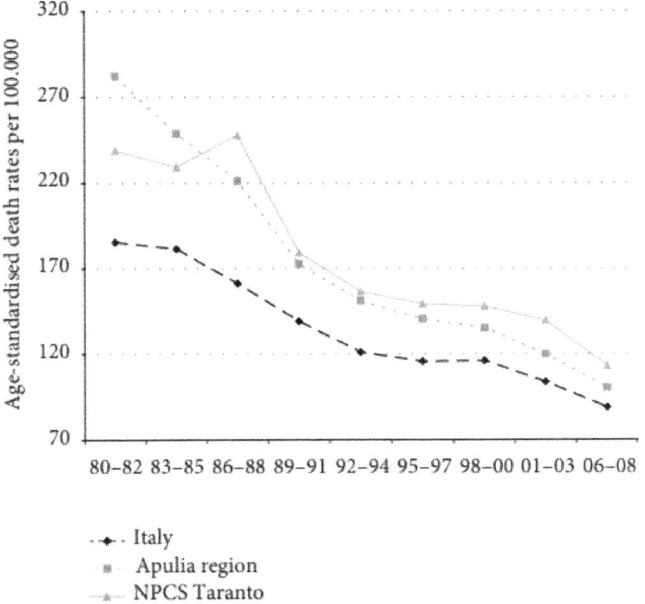

Figure 6: Respiratory diseases. Trends in age-standardised (Italian Census 2001) death rates per 100.000, from selected causes of death (1980–2008) (2004-2005 data were not available). All ages. Men.

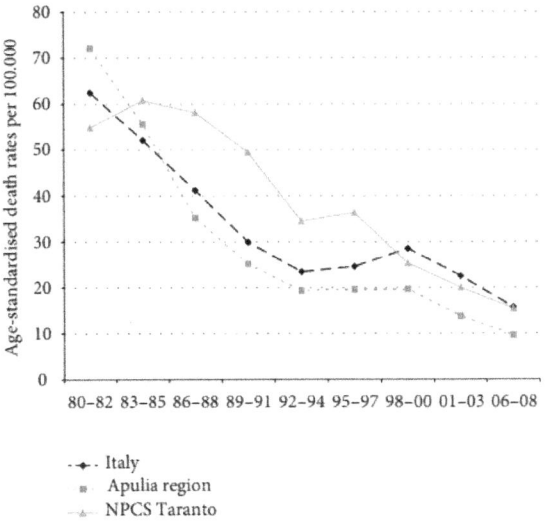

Figure 7: Acute respiratory diseases. Trends in age-standardised (Italian Census 2001) death rates per 100.000, from selected causes of death (1980–2008) (2004-2005 data were not available). All ages. Men.

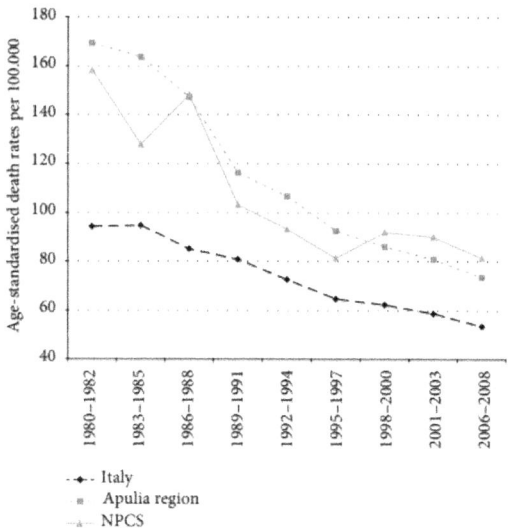

Figure 8: Chronic respiratory diseases. Trends in age-standardised (Italian Census 2001) death rates per 100.000, from selected causes of death (1980–2008) (2004-2005 data were not available). All ages. Men.

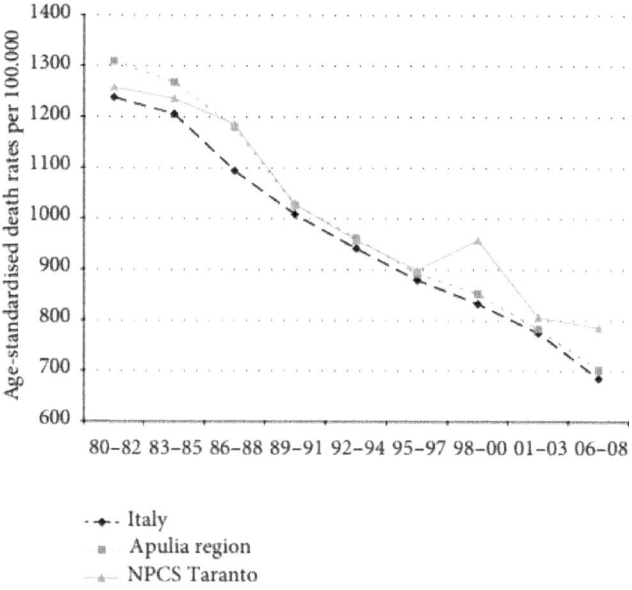

Figure 9: Overall mortality. Trends in age-standardised (Italian Census 2001) death rates per 100.000 (1980–2008) (2004-2005 data were not available). All ages. Women.

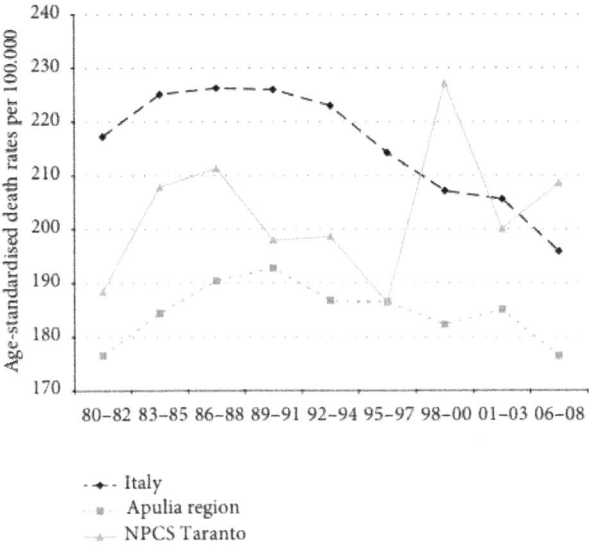

Figure 10: All cancer. Trends in age-standardised (Italian Census 2001) death rates per 100.000, from selected causes of death (1980–2008) (2004-2005 data were not available). All ages. Women.

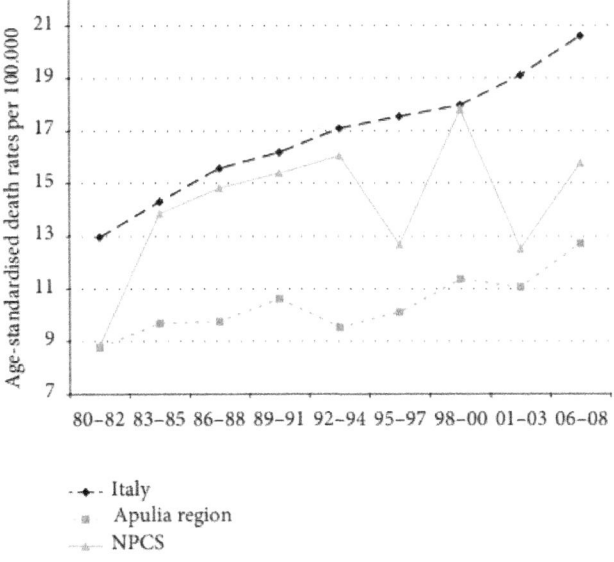

Figure 11: Lung cancers. Trends in age-standardised (Italian Census 2001) death rates per 100.000, from selected causes of death (1980–2008) (2004-2005 data were not available). All ages. Women.

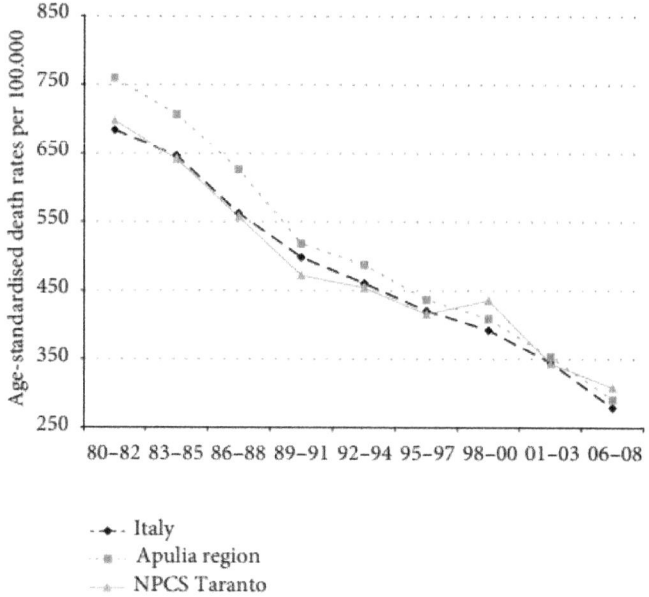

Figure 12: Circulatory diseases. Trends in age-standardised (Italian Census 2001) death rates per 100.000, from selected causes of death (1980–2008) (2004-2005 data were not available). All ages. Women.

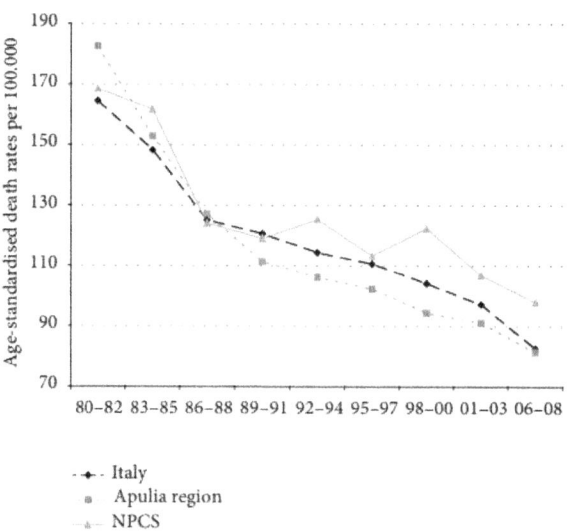

Figure 13: Ischaemic heart diseases. Trends in age-standardised (Italian Census 2001) death rates per 100.000, from selected causes of death (1980–2008) (2004-2005 data were not available). All ages. Women.

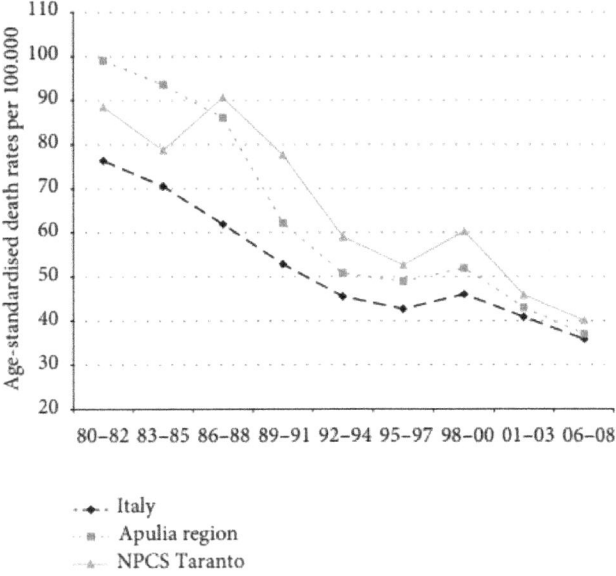

Figure 14: Respiratory diseases. Trends in age-standardised (Italian Census 2001) death rates per 100.000, from selected causes of death (1980–2008) (2004-2005 data were not available). All ages. Women.

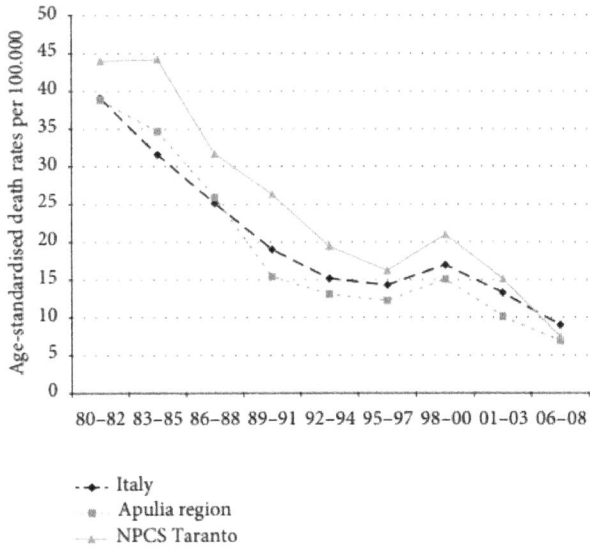

Figure 15: Acute respiratory diseases. Trends in age-standardised (Italian Census 2001) death rates per 100.000, from selected causes of death (1980–2008) (2004-2005 data were not available). All ages. Women.

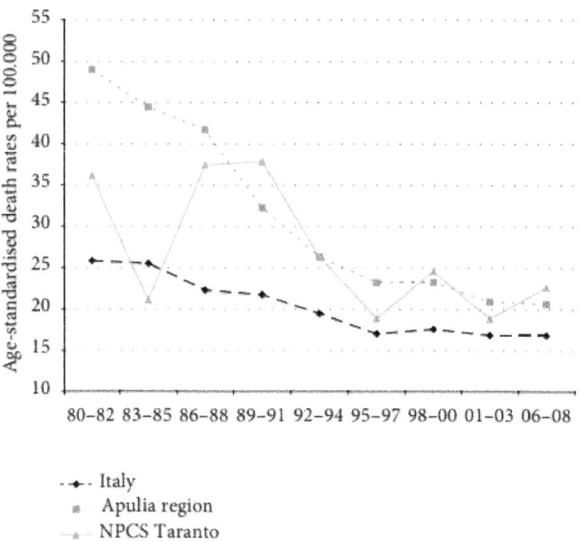

Figure 16: Chronic respiratory diseases. Trends in age-standardised (Italian Census 2001) death rates per 100.000, from selected causes of death (1980–2008) (2004-2005 data were not available). All ages. Women.

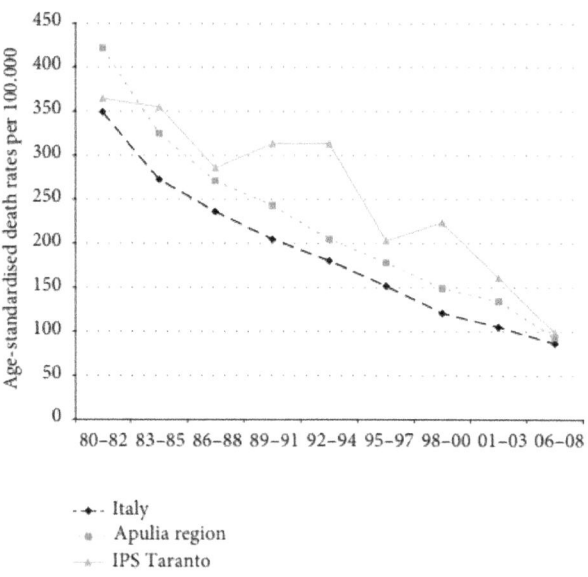

Figure 17: Overall mortality. Trends in age-standardised (Italian Census 2001) death rates per 100.000 (1980–2008) (2004-2005 data were not available). Infant Mortality (0 yrs).

Men

Since many decades, overall mortality in Italy and in Apulia has been declining (resp., 44% and 45%). This favorable trend was also observed in Taranto NPCS, where the regular fall showed a slackening in the last three-year period. Since the early 90s, the SDRs observed in Taranto NPCS were higher than those observed in Apulia, which in turn were lower than the Italian ones; in the most recent three-year period the SDR observed in Taranto site was higher than that in Apulia and Italy (Figure 1). In Italy, mortality from all neoplasms has been falling throughout the study period, while in Taranto NPCS and in Apulia, the trend has been moderately increasing for the same study period. In Taranto NPCS SDRs tend to be higher than in Apulia, which in turn are lower than in Italy (Figure 2).

Mortality from lung cancer has been declining in Italy, Apulia, and Taranto NPCS throughout the years 1980–2008; SDRs in the study site are higher than in Italy and Apulia, since 1995–1997, this differential shows a reduction (Figure 3). Mortality from circulatory diseases in Italy, Apulia, and Taranto site decreased, and the mortality rates almost halved during the study period; yet, in the most recent time, Taranto's mortality rates are higher than in Italy (Figure 4).

Mortality trends from ischemic heart disease have been declining in Italy, Apulia, and Taranto site; yet, since the end of the 1980s, Taranto's SDRs are higher than in Apulia (Figure 5). Mortality trends from respiratory diseases (overall, acute, and chronic) showed a decrease in Italy, Apulia, and Taranto. The rates in Taranto site are higher than those in the Italian ones over the whole study period, the exception being the last decade, when only mortality due to acute respiratory was higher in Taranto (Figures 6, 7, and 8).

Women

Overall mortality among women in Italy, Apulia, and Taranto site showed a long-term decreasing trend: from 1980 to 2008 the decrease was respectively 45%, 46%, and 38%; yet, since the beginning of the 2000s, mortality rates in Taranto are higher compared to those in Apulia and Italy (Figure 9).

Mortality from cancer (all sites) has been decreasing in Italy, remaining stable in Apulia, and increasing in Taranto site (Figure 10).

In contrast with the overall cancer mortality, lung cancer mortality has been rising steadily; in the study period the increase was 59% in Italy, 44% in Apulia, and 78% in Taranto NPCS, where SDRs were higher than in Apulia (Figure 11).

Mortality from circulatory diseases showed a declining trend in Italy, Apulia, and Taranto site (Figure 12). Mortality from ischemic heart disease declined as well, but the rates observed in Taranto NPCS were higher than in Apulia and Italy throughout the study period (Figure 13).

Finally, mortality from respiratory diseases (overall, acute, and chronic) showed a decline in Italy, Apulia, and Taranto site, but the rates in Taranto are higher than in Italy (Figures 14, 15, and 16). The results indicate that some differential between Taranto and other areas are emerging in recent years; it may be attributed to the presence for many years of pollutants in the environment, given that no remediation has been performed in the Taranto area.

Infant Mortality

Infant mortality showed a steady decline both in Italy and Apulia; SDRs in Taranto were decreasing, but they stayed higher in Taranto than in Apulia and Italy (Table S4 and Figure 17).

Cancer Incidence 2006-2007

In Taranto NPCS (Table 4) excesses were observed among both males and females when using for comparison both the rates of South and Islands macroarea and of the province (Taranto and Statte excluded) for all tumors and a number of tumor sites (stomach, colon-rectum, liver, pancreas, lung, mesothelioma, skin melanoma, kidney and other unspecified genitourinary organs, and leukemia). Among males an increase was present for prostate cancer and among females for breast cancer. For most sites, the excesses were confirmed when province rates (Taranto and Statte excluded) were used for reference, thus supporting a major health impact among residents in Taranto NPCS.

Table 4: Cancer incidence Taranto NPCS. Number of observed cases (Obs), standardized incidence ratio crude (SIR); 90% IC: 90% confidence interval; Reference SIR: macroarea South and Islands 2005–2007; Reference SIR: Taranto province excluding NPCS municipalities (TA-NPCS) 2006-2007

Cancer site	Males OBS	Males SIR-macroarea South and Islands (90% CI)	Males SIR (TA-NPCS) (90% CI)	Females OBS	Females SIR-macro-area South and Islands (90% CI)	Females SIR (TA-NPCS) (90% CI)
Head and neck	52	98 (77-123)	131 (103-165)	13	96 (57-153)	134 (79-213)
Stomach	51	130 (102-164)	117 (91-148)	43	167 (127-215)	224 (171-289)
Colon and rectum	145	112 (97-129)	122 (106-140)	131	115 (99-133)	121 (104-140)
Liver	64	110 (88-135)	140 (113-172)	36	110 (82-145)	175 (130-231)
Pancreas	28	100 (71-137)	135 (96-185)	31	113 (82-153)	129 (93-174)
Lung	245	144 (129-160)	150 (135-167)	47	117 (90-149)	148 (114-189)
Skin melanoma	35	214 (158-284)	193 (143-256)	23	143 (98-203)	120 (82-170)
Mesothelioma	21	429 (287-618)	256 (172-369)	3	197 (53-509)	81 (22-209)
Breast	–	–	–	317	130 (118-143)	124 (113-136)
Prostate	204	129 (112-148)	121 (105-139)	–	–	–
Testis	12	109 (63-177)	79 (46-128)	–	–	–
Uterus, cervix	–	–	–	14	93 (56-145)	88 (53-138)
Uterus, body	–	–	–	69	134 (109-164)	188 (152-230)
Ovary	–	–	–	35	119 (88-158)	81 (60-107)
Kidney and other unspecified urinary organs	51	164 (128-207)	201 (157-254)	18	119 (77-176)	114 (74-169)
Bladder	188	141 (125-159)	136 (120-153)	23	62 (42-88)	92 (63-130)
Brain and CNS (malignant)	16	86 (54-131)	88 (55-134)	12	78 (45-126)	65 (37-105)
Thyroid	23	169 (115-239)	126 (86-179)	71	152 (124-185)	94 (76-115)
Hodgkin lymphoma	6	88 (38-174)	63 (27-124)	8	131 (65-236)	70 (33-126)
Non-Hodgkin lymphoma	42	119 (90-154)	160 (122-207)	28	88 (63-121)	143 (102-196)
Myeloma	18	140 (90-208)	135 (87-200)	15	107 (66-165)	97 (60-149)
Leukemia	30	108 (78-146)	82 (59-111)	37	164 (122-216)	103 (77-136)
All tumors excluding skin, nonmalignant brain, and CNS	1338	131 (125-137)	130 (124-136)	1084	126 (120-132)	121 (115-127)

Mortality 1998–2008 and Morbidity 1998–2010 of the Residential Cohort

The study area was divided into 9 districts in Taranto city and 2 municipalities (Massafra and Statte). Figure 18 shows the districts investigated and the location of the industrial area. We considered that the districts located close to the industrial zone were the most areas affected by environmental pollution, especially considering the prevailing winds from northwest. The exposed districts were (1) Tamburi (we also included in this category the small districts of Isola, Porta Napoli, and Lido Azzurro), (2) Borgo, (3) Paolo VI, and (4) Statte. All the other districts were considered the reference zone

(Italia-Montegragnano, San Vito-Lama-Carelli, Salinella, Solito, Corvisea, Talsano, Tre Carrare-Battisti, and Massafra).

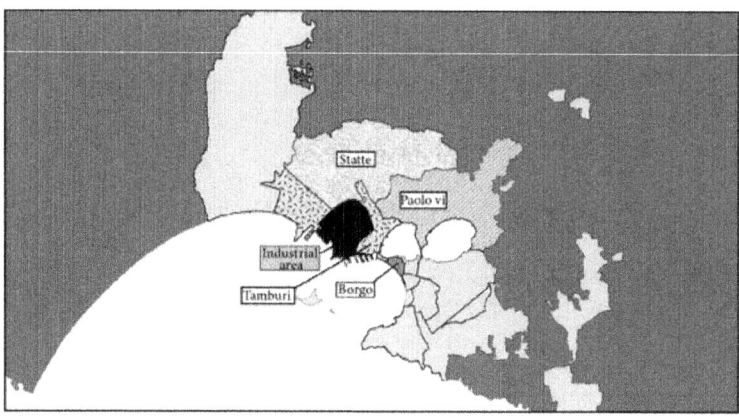

Figure 18: Taranto study area, districts.

A total of 321,356 people (157,031 males and 164,325 females) were enrolled in the cohort. At the time of the enrolment (January 1, 1998), 84.9% of the subjects were already resident in the study area, and 39.1% of them had been residing at the same address for more than 20 years. As far as socioeconomic status is concerned, 35% of the cohort members were in the low SEP category and 21.4% in the high SEP category. The social distribution in the different districts was heterogeneous, with elevated proportion of high social level (62.2%) in some districts in the reference area (San Vito, Lama, Carelli) and low social level in Tamburi (69.4%) and Paolo VI (64.3%). In the Tamburi and Paolo VI districts there was a higher proportion of subjects with previous employment at the steel industry than in other areas.

A total of 3,384,302 person years were accumulated for the cohort. At the end of follow-up (December 31, 2010), 76.6% of the cohort members were alive and resident in the study area, 14.6% had moved, and 28,171 individuals (8.8%) had died. The cause of death was known only for 23,004 individuals deceased by 2008. Only 2.3% of the study subjects were born abroad, and most of the cohort members were born in Taranto (81.6%) and in southern Italy (93.5%).

An analysis of the mortality differences by socioeconomic status (low versus high category) shows (data not in Tables) in both genders a higher mortality for all causes (Hazard ratio—HR 1.25, 95% CI 1.19–1.31 among males and HR 1.18, 95% CI 1.13–1.24 among females), cardiovascular diseases (HR 1.14, 95% CI 1.04–1.25 among males and HR 1.21, 95% CI 1.11–1.32 among females), respiratory diseases (HR 1.89, 95% CI 1.55–2.29 among males and HR 1.38, 95% CI 1.11–1.72 among females), and digestive diseases (HR 1.46, 95% CI 1.19–1.79 among males and HR 1.56, 95% CI 1.24–1.95 among females). Socioeconomic differences in mortality were observed among males for all neoplasms (HR 1.26, 95% CI 1.15–1.37) and cancer of the stomach (HR 1.69, 95% CI 1.10–2.59), larynx (HR 3.32, 95% CI 1.55–7.09), lung (HR 1.40, 95% CI 1.20–1.64), and bladder (HR 1.55, 95% CI 1.08–2.23). The results for hospitalizations confirm the increased risks among subjects in the low socioeconomic category when compared with those in the highest socioeconomic group.

Tables 5 and 6 show, respectively, for males and females, cause-specific mortality in the exposed subareas, that is, Tamburi, Borgo, Paolo VI, and Statte, compared with mortality observed in the reference ones. After adjusting for socioeconomic status, the exposed subareas (Tamburi, Borgo, Paolo VI, and Statte) have a higher mortality for all causes in comparison with the reference in males (in particular, Paolo VI and Tamburi). The most notable increases in mortality among males were in Paolo VI, with 42% excess for all malignant neoplasms (especially lung cancer, +76%), diseases of the cardiovascular (+28%), respiratory (+64%) and digestive (+47%) systems. In Tamburi, an excess was observed among males for all malignant neoplasms (+11%) and cardiovascular diseases (+10%), specifically ischemic heart diseases (+20%). Among females, in Paolo VI, excesses are present for all cancers (+23%), in particular lung, pleural and liver cancer, cardiovascular diseases (+18%), chronic obstructive pulmonary disease (COPD), and digestive system diseases. In Tamburi, excesses were present among females for cardiovascular diseases (+15%), in particular ischemic heart diseases, COPD (+39%), and renal diseases (+57%).

Table 5: Association between district and cause-specific mortality (HR, 90% CI) (males, 1998–2008). Hazard Ratio (HR) from the Cox model stratified by calendar period and adjusted for age (underlying time) and socioeconomic position

Cause of Death (ICD-9-CM)	Reference districts n = 107,909 n	Tamburi n = 14,067			Borgo n = 16,312			Paolo VI n = 10,097			Statte n = 8,283		
		n	HR	90% CI	n	HR	90% CI	n	HR	90% CI	n	HR	90% CI
All causes (001-999)*	9,378	1,470	1,12	1,07 1,18	1,973	1,07	1,03 1,12	684	1,27	1,19 1,36	654	1,08	1,01 1,15
Malignant cancers (140-208)	2,650	400	1,11	1,01 1,21	505	1,00	0,93 1,09	223	1,42	1,26 1,59	178	1,05	0,93 1,20
Stomach (151)	126	22	1,24	0,83 1,84	28	1,20	0,85 1,71	12	1,62	0,97 2,69	7	0,85	0,45 1,61
Colorectal (153-154)	220	19	0,62	0,41 0,92	48	1,13	0,87 1,48	12	1,07	0,65 1,75	11	0,79	0,48 1,32
Trachea, bronchus, and lung (162)	829	127	1,09	0,93 1,28	150	0,97	0,84 1,13	94	1,76	1,46 2,11	61	1,12	0,90 1,39
Pleura (163)	80	12	1,09	0,64 1,85	16	1,08	0,68 1,71	6	1,19	0,59 2,42	2	0,39	0,12 1,27
Prostate (185)	244	45	1,42	1,07 1,89	53	0,98	0,76 1,26	8	0,84	0,46 1,52	12	0,85	0,52 1,39
Bladder (188)	189	34	1,20	0,87 1,66	29	0,73	0,53 1,02	13	1,45	0,90 2,35	13	1,17	0,73 1,87
Kidney (189)	21	5	2,23	0,93 5,33	5	1,30	0,56 2,98	2	1,85	0,53 6,39	5	3,69	1,62 8,45
Brain and other parts of CNS (191-192; 225)	83	15	1,37	0,85 2,23	18	1,23	0,80 1,90	10	1,64	0,93 2,88	6	1,07	0,53 2,14
Lymphatic and hematopoietic tissue (200-208)	212	27	1,05	0,74 1,49	41	1,07	0,81 1,43	14	1,01	0,64 1,60	9	0,69	0,39 1,20
Neurological diseases (330-349)	190	28	1,09	0,77 1,54	30	0,76	0,54 1,05	7	0,72	0,38 1,36	15	1,32	0,85 2,06
Cardiovascular diseases (390-459)	2,442	378	1,10	1,00 1,21	551	1,02	0,94 1,10	147	1,28	1,11 1,47	137	0,93	0,81 1,08
Cardiac diseases (390-429)	1,688	260	1,09	0,98 1,23	387	1,03	0,94 1,13	106	1,27	1,08 1,51	84	0,82	0,68 0,99
Ischemic heart diseases (410-414)	733	116	1,20	1,01 1,42	152	1,04	0,89 1,20	56	1,37	1,09 1,73	30	0,66	0,49 0,90
Cerebrovascular diseases (430-438)	551	86	1,06	0,87 1,29	109	0,87	0,73 1,04	25	1,07	0,76 1,50	43	1,34	1,04 1,75
Respiratory diseases (460-519)	697	122	1,08	0,91 1,27	177	1,05	0,91 1,20	48	1,64	1,28 2,11	59	1,46	1,17 1,82
COPD (490-492, 494, 496)	469	93	1,17	0,96 1,41	110	0,94	0,79 1,12	32	1,70	1,25 2,32	39	1,44	1,10 1,90
Diseases of the digestive system (520-579)	527	81	1,06	0,86 1,30	111	1,07	0,90 1,27	47	1,47	1,14 1,90	27	0,79	0,57 1,09
Renal diseases (580-599)	146	27	1,36	0,94 1,95	37	1,13	0,83 1,54	3	0,50	0,19 1,30	9	1,00	0,57 1,76

*Referring to the period 1998-2010.

Table 6: Association between district and cause-specific mortality (HR, 90% CI) (females, 1998–2008). Hazard Ratio (HR) from the Cox model stratified by calendar period and adjusted for age (underlying time) and socioeconomic position

Cause of death (ICD-9-CM)	Reference districts n = 112,897 n	Tamburi n = 14,625			Borgo n = 18,528			Paolo VI n = 9,714			Statte n = 8,271		
		n	HR	90% CI	n	HR	90% CI	n	HR	90% CI	n	HR	90% CI
All causes (001–999)*	9.015	1.479	1,09	1,04 1,15	2 482	1,01	0,97 1,05	489	1,28	1,18 1,38	547	1,06	0,98 1,14
Malignant cancers (140-208)	1.900	230	0,84	0,75 0,95	434	0,95	0,87 1,04	126	1,23	1,06 1,44	102	0,92	0,78 1,08
Stomach (151)	96	20	1,52	0,99 2,34	24	1,01	0,69 1,48	7	1,47	0,76 2,83	7	1,31	0,69 2,51
Colorectal (153-154)	226	23	0,62	0,43 0,90	45	0,78	0,59 1,02	16	1,35	0,87 2,08	7	0,54	0,29 1,02
Trachea, bronchus and lung (162)	144	15	0,76	0,48 1,20	34	1,06	0,77 1,46	13	1,71	1,05 2,79	6	0,68	0,34 1,34
Pleura (163)	20	2	0,66	0,19 2,31	6	1,16	0,53 2,52	3	2,95	1,03 8,46	0		
Breast (174)	349	41	0,92	0,69 1,22	89	1,18	0,96 1,44	28	1,29	0,93 1,79	22	1,04	0,72 1,49
Bladder (188)	33	7	1,23	0,61 2,50	12	1,13	0,64 1,97	2	1,29	0,39 4,34	1	0,58	0,11 3,06
Kidney (189)	17	0			3	0,87	0,30 2,49	0	0,00	0,00 0,00	1	1,06	0,19 5,79
Brain and other parts of CNS (191-192; 225)	90	6	0,48	0,24 0,97	17	0,85	0,55 1,33	4	0,67	0,29 1,57	7	1,30	0,68 2,48
Lymphatic and hematopoietic tissue (200-208)	202	22	0,74	0,50 1,08	33	0,65	0,47 0,88	11	0,98	0,59 1,65	11	0,99	0,60 1,66
Neurological diseases (330-349)	216	35	1,08	0,79 1,47	50	0,83	0,64 1,08	13	1,68	1,04 2,71	11	0,87	0,52 1,45
Cardiovascular diseases (390-459)	2.945	529	1,15	1,06 1,24	876	0,93	0,88 1,00	125	1,18	1,01 1,37	166	0,98	0,86 1,12
Cardiac diseases (390-429)	1.910	371	1,24	1,12 1,37	623	1,04	0,96 1,12	84	1,22	1,01 1,47	90	0,81	0,68 0,97
Ischemic heart diseases (410-414)	565	124	1,46	1,23 1,73	171	1,02	0,88 1,18	24	1,15	0,81 1,63	27	0,86	0,62 1,19
Cerebrovascular diseases (430-438)	820	122	0,93	0,79 1,10	207	0,77	0,68 0,88	35	1,19	0,90 1,59	62	1,38	1,11 1,71
Respiratory diseases (460-519)	476	82	1,09	0,89 1,34	169	1,09	0,94 1,26	22	1,26	0,88 1,82	34	1,28	0,95 1,71
COPD (490-492, 494, 496)	220	49	1,39	1,06 1,83	70	0,97	0,77 1,21	16	2,14	1,38 3,29	14	1,16	0,74 1,83
Diseases of the digestive system (520-579)	484	77	0,95	0,77 1,16	119	0,88	0,74 1,04	29	1,43	1,04 1,97	30	1,13	0,83 1,54
Renal diseases (580-599)	166	38	1,57	1,15 2,14	49	1,01	0,77 1,33	10	1,68	0,98 2,90	11	1,12	0,67 1,87

*Referring to the period 1998-2010.

Tables 7 and 8 display the results for hospital admissions. The hospitalization analysis confirms the mortality results, documenting the major health impact on residents in Tamburi and Paolo VI areas, where excesses were observed among males for a number of causes such as lung cancer (29% and 61% in Tamburi and Paolo VI, resp.), neurological (26% and 43% in Tamburi and Paolo VI, resp.), cardiovascular (18% and 32% in Tamburi and Paolo VI, resp.), respiratory (36% and 52% in Tamburi and Paolo VI, resp.), and renal diseases (35% both in Tamburi and Paolo VI). Among females, similar excesses were observed for cardiovascular diseases (15 and 31% in Tamburi and Paolo VI, resp.), respiratory diseases (28% and 39% in Tamburi and Paolo VI, resp.), digestive diseases (18% and 25% in Tamburi and Paolo VI, resp.), and renal diseases (47% and 35% among males in Tamburi and Paolo VI, resp.). In Paolo VI, pleural (235%) and breast cancer (33%) were also in excess among females.

Table 7: Association between district and cause-specific hospitalization (HR, 90% CI) (males, 1998–2010). Hazard Ratio (HR) from the Cox model stratified by calendar period and adjusted for age (underlying time) and socioeconomic position

Diagnosis (ICD-9-CM)	Reference districts n = 108,272 n	Tamburi n = 14,067			Borgo n = 16,312			Paolo VI n = 10,097			Statte n = 8,283		
		n	HR	90% CI	n	HR	90% CI	n	HR	90% CI	n	HR	90% CI
Malignant cancers (140-208)	4.818	685	1,12	1,05 1,21	861	1,06	0,99 1,12	444	1,31	1,21 1,43	354	1,06	0,97 1,16
Stomach (151)	166	27	1,21	0,85 1,73	30	1,05	0,75 1,46	19	1,63	1,09 2,46	15	1,29	0,83 2,02
Colorectal (153-154)	520	52	0,86	0,67 1,10	73	0,87	0,70 1,07	28	0,82	0,59 1,13	39	1,08	0,82 1,41
Trachea, bronchus, and lung (162)	866	149	1,29	1,10 1,50	156	1,06	0,92 1,22	101	1,61	1,34 1,92	60	0,98	0,78 1,22
Pleura (163)	75	18	1,80	1,14 2,84	18	1,38	0,89 2,15	7	1,44	0,74 2,79	4	0,77	0,33 1,79
Connective and other soft tissue (171)	32	4	0,91	0,37 2,24	10	1,80	0,98 3,31	5	1,66	0,74 3,72	3	1,30	0,48 3,52
Prostate (185)	639	80	1,10	0,90 1,35	113	1,04	0,88 1,24	40	0,98	0,75 1,29	51	1,22	0,96 1,56
Testis (186)	49	3	0,42	0,15 1,13	9	1,23	0,67 2,25	2	0,40	0,12 1,34	5	1,31	0,60 2,84
Bladder (188)	787	106	1,12	0,94 1,34	121	0,90	0,77 1,06	84	1,62	1,34 1,97	50	0,95	0,74 1,20
Kidney (189)	149	23	1,26	0,85 1,85	37	1,47	1,08 1,99	15	1,41	0,89 2,22	14	1,37	0,86 2,17
Brain and other parts of CNS (191-192; 225)	187	23	0,98	0,67 1,43	30	1,01	0,73 1,41	14	0,89	0,56 1,41	16	1,19	0,77 1,83
Lymphatic and hematopoietic tissue (200-208)	400	58	1,20	0,94 1,53	80	1,26	1,03 1,55	35	1,13	0,84 1,52	24	0,84	0,60 1,19
Neurological diseases (330-349)	1.850	329	1,26	1,14 1,40	337	1,11	1,00 1,22	226	1,43	1,27 1,61	140	1,04	0,90 1,20
Cardiovascular diseases (390-459)	14.504	2.078	1,18	1,13 1,23	2.457	1,03	1,00 1,07	1.388	1,32	1,26 1,39	1.076	1,06	1,01 1,12
Cardiac diseases (390-429)	9.866	1.411	1,20	1,15 1,27	1.699	1,04	0,99 1,09	957	1,41	1,33 1,49	752	1,10	1,04 1,17
Acute coronary events (410-411)	2.328	310	1,13	1,02 1,26	396	1,09	0,99 1,19	241	1,39	1,24 1,56	167	1,00	0,87 1,14
Heart failure (428)	1.878	317	1,21	1,09 1,34	375	1,03	0,94 1,14	180	1,54	1,35 1,76	119	0,97	0,83 1,13
Cerebrovascular diseases (430-438)	3.124	525	1,30	1,19 1,41	581	1,04	0,96 1,12	211	1,01	0,89 1,13	226	1,09	0,97 1,22
Respiratory diseases (460-519)	8.906	1.836	1,36	1,30 1,43	1.493	1,01	0,96 1,06	1.255	1,52	1,44 1,60	752	1,12	1,05 1,19
Acute respiratory infections (460-466, 480-487)	3.769	961	1,51	1,42 1,61	746	1,19	1,11 1,27	625	1,54	1,44 1,66	315	1,05	0,95 1,15
COPD (490-492, 494, 496)	2.350	450	1,31	1,20 1,43	379	0,85	0,78 0,93	252	1,60	1,43 1,79	203	1,29	1,14 1,46
Diseases of the digestive system (520-579)	15.628	2.465	1,20	1,15 1,24	2.301	0,94	0,91 0,98	1.682	1,23	1,18 1,29	1.229	1,06	1,01 1,12
Renal diseases (580-599)	3.252	562	1,35	1,25 1,46	521	0,98	0,90 1,06	331	1,35	1,22 1,49	282	1,22	1,10 1,36

Table 8: Association between district and cause-specific hospitalization (HR, 90% CI) (females, 1998–2010). Hazard Ratio (HR) from the Cox model stratified by calendar period and adjusted for age (underlying time) and socioeconomic position

Diagnosis (ICD-9-CM)	Reference districts n = 113,187	Tamburi n = 14,625			Borgo n = 18,528			Paolo VI n = 9,714			Statte n = 8,271		
	n	n	HR	90% CI	n	HR	90% CI	n	HR	90% CI	n	HR	90% CI
Malignant cancers (140-208)	3.878	530	1,03	0,95 1,12	713	0,91	0,85 0,97	295	1,17	1,06 1,29	225	0,91	0,81 1,02
Stomach (151)	129	23	1,26	0,85 1,85	24	0,89	0,61 1,28	7	1,06	0,55 2,02	9	1,10	0,62 1,94
Colorectal (153-154)	483	60	0,93	0,74 1,18	62	0,62	0,50 0,78	35	1,21	0,90 1,62	17	0,58	0,39 0,88
Trachea, bronchus, and lung (162)	149	17	0,89	0,57 1,37	30	0,94	0,67 1,33	10	1,04	0,60 1,80	5	0,53	0,25 1,12
Pleura (163)	23	3	0,87	0,31 2,44	2	0,44	0,13 1,48	4	3,35	1,31 8,54	2	1,25	0,37 4,22
Breast (174)	990	127	1,03	0,87 1,21	179	0,98	0,86 1,12	94	1,33	1,11 1,59	59	0,89	0,72 1,11
Vescica (188)	146	22	1,12	0,76 1,66	37	1,16	0,85 1,58	5	0,61	0,29 1,29	5	0,56	0,27 1,19
Kidney (189)	78	10	0,95	0,53 1,68	13	0,84	0,51 1,39	5	1,06	0,49 2,28	10	2,10	1,21 3,66
Brain and other parts of CNS (191-192; 225)	204	25	0,94	0,65 1,35	30	0,77	0,55 1,06	17	1,18	0,77 1,80	17	1,27	0,84 1,93
Lymphatic and hematopoietic tissue (200-208)	398	55	1,03	0,80 1,31	58	0,72	0,57 0,91	26	1,00	0,71 1,40	24	0,95	0,67 1,35
Neurological diseases (330-349)	2.151	351	1,11	1,01 1,23	378	0,89	0,81 0,98	168	1,06	0,93 1,21	141	0,98	0,85 1,14
Cardiovascular diseases (390-459)	13.500	2.072	1,15	1,11 1,20	2.611	0,90	0,87 0,93	1.059	1,31	1,25 1,39	888	1,05	0,99 1,11
Cardiac diseases (390-429)	9.366	1.478	1,17	1,12 1,23	1.897	0,93	0,89 0,97	752	1,40	1,31 1,49	632	1,09	1,02 1,17
Acute coronary events (410-411)	1.060	197	1,32	1,16 1,51	254	1,08	0,96 1,22	88	1,42	1,18 1,72	64	1,02	0,82 1,26
Heart failure (428)	2.449	368	0,95	0,87 1,05	628	1,02	0,94 1,09	154	1,32	1,15 1,52	106	0,77	0,65 0,90
Cerebrovascular diseases (430-438)	3.595	600	1,15	1,06 1,24	728	0,84	0,79 0,90	220	1,21	1,07 1,35	212	1,01	0,90 1,13
Respiratory diseases (460-519)	6.673	1.336	1,28	1,21 1,34	1.273	1,00	0,95 1,05	813	1,39	1,31 1,49	514	1,07	0,99 1,16
Acute respiratory infections (460-466, 480-487)	3.020	705	1,39	1,29 1,49	599	1,08	1,00 1,17	422	1,37	1,25 1,49	228	0,98	0,88 1,10
COPD (490-492, 494, 496)	1.433	262	1,19	1,06 1,33	325	0,94	0,85 1,04	126	1,62	1,39 1,89	91	1,09	0,91 1,30
Diseases of the digestive system (520-579)	12.952	2.067	1,18	1,13 1,22	2.038	0,89	0,85 0,92	1.288	1,25	1,19 1,31	905	1,00	0,95 1,06
Renal diseases (580-599)	3.187	662	1,47	1,37 1,59	609	0,99	0,92 1,07	320	1,35	1,22 1,49	248	1,17	1,05 1,31

DISCUSSION

The health impact of residence in Taranto NPCS was investigated using different epidemiological approaches: geographical (mortality and cancer incidence at municipality level), historical (mortality time trends at municipality level), and residential cohort studies (mortality and hospital discharge records at individual level). We adopted an analysis at small-area scale which includes a mix of small area and individual based data. The design issues relevant to this type of investigation have been thoroughly examined by Elliott and Savitz [27], some of them are briefly examined in the following paragraphs.

In environmental epidemiology, exposure ascertainment is a key phase because the exposure/s affecting the study population should ideally be described in detail, while in most instances the available exposure information is indirect and qualitative. In ecological investigations, the exposure/s can be a single event from a point emission source of some contaminants; more often the contaminants are heterogeneous mixture progressively polluting different matrices in the area. For example, in SENTIERI Project the sources ofenvironmental exposures were abstracted from the legislative decrees defining sites' boundaries and fixed on the basis of the possible sources of contamination (e.g., chemical industry, steel plants, and landfills). A further limitation lies in the implicit assumption that all residents in the area under investigation experience the same exposures, while exposure variability is likely to be substantial, due to many factors (e.g., concentration of contaminants and their diffusion to soil and water, distance of residence from polluting sources). The possible consequences of such nondifferential exposure misclassification are complex, and the direction of the resulting bias is not predictable [28]. In addition, information exposure source/s with possible health impact, such as concurrent air pollution from road traffic and exposures in the occupational setting, are often not available. Finally, vital statistics are accessible for a given administrative area whose boundaries hardly correspond to the distribution of environmental pollutants, so that the misclassification of exposure (and loss of statistical power) is common. A more detailed description of these limitations of ecological study design is available [15, 28, 29].

Exposure ascertainment is a critical issue in ecological investigations as well as in studies based on individuals, as cohort of residents are. In this case residential history, overtime information on exposure/s in different residences and different environmental matrices, daily variability, and seasonal variability should be available to classify subjects in different exposure categories. Obtaining individual-based measures of exposure as described above is clearly infeasible, and modeling of exposures, ranging from simple measures such as distance from a point source or distance to nearest road to more complex estimation, for example, dispersion modeling around a point source, is used. The possible exposure misclassification in such approaches has been discussed in [27].

As far as outcome measures are concerned, many studies of environmental healthin polluted areas consider mortality, based on death records. However, the analysis of hospital discharge records, ad hoc registry data of specific pathologies (e.g., cancer, congenital malformations) can give a better picture of the health profile of residents in NPCSs. The key issue is that, whatever the outcome under investigation, databases need to be validated for use in epidemiological studies.

Reporting the event of death is usually exhaustive; therefore the overall mortality can be analyzed with confidence [30]. In Italy, validity of cause of death certification has been documented for specific diseases [31–34]. The validity of hospital discharge records (HDRs), indicators of hospital activity, has not been systematically evaluated, although in Italy some critical aspects of this novel utilization of HDRs have been examined [35].

Another crucial aspect in environmental health studies is that factors such as socioeconomic status, occupational exposures/s, and individual lifestyles can have an etiologic role on the health effects under study thus possibly confounding the exposure-disease relationships.

For a review of adjustment for socioeconomic status using census data in ecological studies of environment and health refer to Pasetto et al, 2010 [36]. To account for deprivation in SENTIERI Project, mortality data both crude and adjusted were analyzed using an ad hoc built deprivation index [25]. Also in individual-based studies, the analyses can be adjusted for socioeconomic factors, usually with an aggregate indicator based on residence address. In the present cohort of residents in Taranto NPCS SEP was assigned to each participant on the basis of the geocoded addresses at the beginning of the follow-up.

Occupational exposure/s are also potential confounders in environmental health studies. The ecological components of the present investigation are affected by this limitation, while for the cohort study individual occupational history was traced through the national insurance company (INPS) database, and the subcohort of individuals employed in industries located in the area was identified indicating a high proportion of past employment at the steel plant among residents in Tamburi and Paolo VI.

Again on this topic, with specific reference to pleural mesothelioma, it should be noted that the Italian National Mesothelioma Registry analysis of residential asbestos exposure showed that steel mills and iron foundries were the second most frequent sources of asbestos in the neighborhood (after asbestos-cement factories), ranking equal to asbestos textiles production [37]. Furthermore a case-control study in the area of Taranto (HDR, 1998–2002) adjusted for occupational exposures [11], showed an increase in malignant pleural neoplasms among residents close to the steel mill (or 1.62, 95% CI 0.37–7.10, 11 cases) and the coke plant (or 2.18, 95% CI 0.31–15.31, 9 cases).

Another aspect to consider in environmental epidemiology is that large populations are needed to study many of the health concerns of greatest interest. In ecological investigations, the reference populations should be selected balancing the need for comparability of study and reference populations for factors other than the environmental exposure/s with possible health impact (socioeconomic status and lifestyle factors as diet and tobacco use) and the requirement of sufficiently numerous populations to have stable reference rates also for rare diseases. These needs are satisfied in the present investigation where national, macroregional, and regional populations were used for comparison in the mortality, time trend, and cancer incidence analysis. In the cohort study an internal comparison was carried out, and the reference population was composed of residents in districts distant from the industrial area.

For chronic diseases including most cancers, latency effects are important, such that exposures experienced many years previously, or accumulated exposures, may be crucial. In Taranto NPCS, the mortality (1995–2009) and the time trends analyses (1980–2008) show consistent results and cover a time span which should encompass latency effects. Analogously in the cohort of residents, most subjects were present at enrolment in 1998 (85%), and half of them had a residence duration of 10 or more years.

A brief comment on the use of 90% CI in our analyses is needed. In this respect we refer to Sterne and Smith [38], who affirm that confidence intervals for the main results should always be included, but 90% rather than 95% levels should be used. CI should not be used as a surrogate means of examining significance at the conventional 5% level. Interpretation of CI should focus on the implications (clinical importance) of the range of values in the interval.

CONCLUSIONS

In Taranto NPCS, mortality data at municipality level analyzed, in the context of SENTIERI Project, time trend analysis and cancer incidence results coherently showed, in both genders, excess risks for a number of causes of death, among them: all causes, all cancers, lung cancer, cardiovascular, and respiratory diseases, both acute and chronic. For these causes, an etiologic role of environmental exposure present in Taranto NPCS can be supported on the basis of a priori evaluation of the epidemiological evidence completed in SENTIERI. In the cohort study among residents in the districts nearer to the industrial area, excess mortality/morbidity risks were shown for natural cause, cancers, cardiovascular, and respiratory diseases. These excesses were also observed in low socioeconomic position groups compared to high ones, some of them could be explained on the basis of previous employment of residents in industries active in the study area.

As discussed above the present results for Taranto NPCS are based on sound study design and valid data, which make a low potential for bias and help to strengthen etiologic inference. The present findings further corroborate the need to promptly proceed with environmental cleanup interventions.

It should not be disregarded the fact that most diseases showing an increased risk have multifactorial etiology, therefore interventions of proven efficacy, such as smoking cessation, food education, measures for cardiovascular risk reduction, and breast cancer screening programs, should be planned. To build a climate of confidence and trust between citizens and public institutions, study results and public health actions are to be communicated objectively and transparently.

APPENDICES

A. International Classification of Diseases to Code Mortality: From ICD-9 to ICD-10

Mortality data are coded according to the International Classification of Disease (ICD) which has been revised approximately every 10 years; the purpose of the revision is to stay abreast of medical advances in terms of disease nomenclature and etiology. In Italy, deaths have been coded according to the Ninth Revision (ICD-9) until 2002 [39]; since 2003, the Tenth Revision (ICD-10) has been adopted [40].

ICD-10 differs from ICD-9 in several respects: ICD-10 is far more detailed than ICD-9, with about 12,000 categories versus about 5,000 categories; ICD-10 uses alphanumeric codes compared with numeric codes in ICD-9; some additions and modifications were made to the chapters in the ICD; and some of the coding rules and rules for selecting the underlying cause of death have been changed.

Because of these modifications, comparability studies, also called bridge-coding studies, were carried out to measure the effects of a new revision of the ICD on the comparability with the previous revision of mortality statistics by cause of death. These studies involve the dual classification of a single-year mortality data, that is, classifying the underlying cause of death on mortality records by both the new revision and the previous revision. The key element of a comparability study is the "comparability ratio", which is derived from the dual classification.

Operationally, the comparability ratio for the cause of death i (C_i) is calculated as follows:

$$Ci = \frac{D_{i,\text{ICD-10}}}{D_{i,\text{ICD-9}}}, \quad (A.1)$$

where $D_{i,\text{ICD-10}}$ is the number of deaths due to cause i classified by ICD-10. $D_{i,\text{ICD-9}}$, is the number of deaths due to cause i classified by ICD-9.

A comparability ratio of 1.00 indicates that the same number of deaths was assigned to cause i under both ICD-9 and ICD-10.

Comparability studies between ICD-9 and ICD-10 have been conducted in the USA, in Europe (by Eurostat), and also in Italy (by the Italian Census Bureau, ISTAT) [41–44]. The Italian study [44] documented that comparability ratios of the main causes of death are close to the value of one.

Table S1 presents the comparability study results that referred to some causes investigated in the time trend analysis.

In Italy the definition of municipalities has varied over time: some municipalities have been suppressed, others have been modified, and others have been created.

Generally, these modifications are not implemented at the same time by the office that registers deaths (Civil Status Office) and by the office that registers the population (General Registry office).

Therefore, it may occur that a "new" municipality (created by the subdivision of another municipality) creates first the General Registry, beginning to register its own populations, while the deaths continue to be registered by the Civil Status Office of the "old" municipality [45, 46].

In the context of SENTIERI Project the situation of all of the municipalities constituting the NPCSs has been studied [47].

As far as Taranto and Statte municipalities, that constitute Taranto NPCS, the description is as follows: until 1993, Statte was a district of Taranto; in 1994 it became a new municipality and created immediately its own General Registry Office; on the other hand, its own Civil Status office was set up some years later (1998).

To "synchronize" data regarding deaths and populations of both Statte and Taranto, we decided to attribute the population and the deaths of Statte to Taranto until 1997; since 1998, they have been attributed to Statte itself.

B. Mortality Trends from Selected Causes in Taranto NPCS, 1980–2008

See Supplementary Tables S2, S3, and S4.

ACKNOWLEDGMENTS

SENTIERI Project was funded, for the years 2008–2010, by the Italian Ministry of Health (Strategic Programme Environment and Health) and for the years 2009–2013 by the Ministry of Health's Project CCM 2009 "Epidemiological surveillance of populations living in contaminated sites". The authors wish to thank Letizia Sampaolo for the editorial assistance and language revision.

REFERENCES

1. European Environment Agency, "Progress in management of contaminated sites (CSI 015): assessment published Aug 2007," http://www.eea.europa.eu/data-and-maps/indicators/progress-in-management-of-contaminated-sites/progress-in-management-of-contaminated-1.
2. L. Liberti, M. Notarnicola, R. Primerano, and G. Vitucci, "Air pollution from a large steel factory: toxic contaminants from coke-ovenplants," in Proceedings of the 12th international conference on modelling, monitoring and management of air pollution, pp. 485–494, WIT Press, Southampton, UK, 2004.
3. M. M. Storelli and G. O. Marcotrigiano, "Bioindicator organisms: heavy metal pollution evaluation in the Ionian Sea (Mediterranean Sea, Italy)," Environmental Monitoring and Assessment, vol. 102, no. 1–3, pp. 159–166, 2005.
4. L. Liberti, M. Notarnicola, R. Primerano, and P. Zannetti, "Air pollution from a large steel factory: polycyclic aromatic hydrocarbon emissions from coke-oven batteries," Journal of the Air and Waste Management Association, vol. 56, no. 3, pp. 255–260, 2006.
5. C. Gariazzo, V. Papaleo, A. Pelliccioni, G. Calori, P. Radice, and G. Tinarelli, "Application of a Lagrangian particle model to assess the impact of harbour, industrial and urban activities on air quality in the Taranto area, Italy," Atmospheric Environment, vol. 41, no. 30, pp. 6432–6444, 2007.
6. L. Bisceglia, R. Giua, A. Morabito et al., "Source apportionment of benzo(a)pyrene in Taranto and carcinogenic risk estimate in general population," Giornale Italiano di Medicina del Lavoro ed Ergonomia, vol. 32, no. 4, supplement, pp. 355–356, 2010 (Italian).
7. A. Di Leo, N. Cardellicchio, S. Giandomenico, and L. Spada, "Mercury and methylmercury contamination in Mytilus galloprovincialis from Taranto Gulf (Ionian Sea, Southern Italy): risk evaluation for consumers," Food and Chemical Toxicology, vol. 48, no. 11, pp. 3131–3136, 2010.
8. S. Giandomenico, L. Spada, C. Annicchiarico, G. Assennato, N. Cardellicchio, N. Ungaro, et al., "Chlorinated compounds and polybrominated diphenyl ethers (PBDEs) in mussels (Mytilus

galloprovincialis) collected from Apulia Region coasts," Marine Pollution Bulletin, vol. 73, no. 1, pp. 243–251, 2013.

9. I. Iavarone, E. De Felip, A. M. Ingelido, N. Iacovella, A. Abballe, S. Valentini, et al., "Exploratory biomonitoring study among workers of livestock farms of the Taranto Province," Epidemiologia and Prevenzione, vol. 36, no. 6, pp. 321–331, 2012 (Italian).

10. M. Martuzzi, F. Mitis, A. Biggeri, B. Terracini, and R. Bertollini, "Environment and health status of the population in areas with high risk of environmental crisis in Italy," Epidemiologia and prevenzione, vol. 26, no. 6, supplement, pp. 1–53, 2002 (Italian).

11. A. Marinaccio, S. Belli, A. Binazzi et al., "Residential proximity to industrial sites in the area of Taranto (Southern Italy): a case-control cancer incidence study," Annali dell›Istituto Superiore di Sanita, vol. 47, no. 2, pp. 192–199, 2011.

12. P. Comba, R. Pirastu, S. Conti, M. De Santis, I. Iavarone, G. Marsili, et al., "Environment and health in Taranto, Southern Italy: epidemiological studies and public health recommendations," Epidemiologia and Prevenzione, vol. 36, no. 6, pp. 305–320, 2012 (Italian).

13. F. Mataloni, M. Stafoggia, E. Alessandrini, M. Triassi, A. Biggeri, and F. Forastiere, "A cohort study on mortality and morbidity in the area of Taranto, Southern Italy," Epidemiologia and Prevenzione, vol. 36, no. 5, pp. 237–252, 2012 (Italian).

14. R. Pirastu, C. Ancona, I. Iavarone et al., "SENTIERI Project. Mortality study of residents in Italian polluted sites: evaluation of the epidemiological evidence," Epidemiologia and prevenzione, vol. 34, no. 5-6, pp. 1–2, 2010 (Italian).

15. R. Pirastu, R. Pasetto, A. Zona, C. Ancona, I. Iavarone, M. Martuzzi, et al., "The health profile of population living in contaminated sites: SENTIERI approach," Journal of Environmental and Public Health, vol. 2013, Article ID 939267, 13 pages, 2013.

16. R. Pirastu, C. Ancona, I. Iavarone et al., "SENTIERI Project. Mortality study of residents in Italian polluted sites: evaluation of the epidemiological evidence," Epidemiologia and prevenzione, vol. 34, supplement 3, no. 5-6, pp. 1–2, 2010 (Italian).

17. C. A. Pope III, "Respiratory hospital admissions associated with PM10 pollution in Utah, Salt Lake, and Cache Valleys," Archives of Environmental Health, vol. 46, no. 2, pp. 90–97, 1991.

18. R. S. Bhopal, S. Moffatt, T. Pless-Mulloli et al., "Does living near a constellation of petrochemical, steel, and other industries impair

health?" Occupational and Environmental Medicine, vol. 55, no. 12, pp. 812–822, 1998.

19. P. R. Lewis, M. J. Hensley, J. Wlodarczyk et al., "Outdoor air pollution and children›s respiratory symptoms in the steel cities of New South Wales," Medical Journal of Australia, vol. 169, no. 9, pp. 459–463, 1998.

20. J. Petrela, V. M. Câmara, G. Kennedy, B. Bouyahi, and J. Zayed, "Health effects of residential exposure to aluminum plant air pollution," Archives of Environmental Health, vol. 56, no. 5, pp. 456–460, 2001.

21. M. Wilhelm, G. Eberwein, J. Hölzer et al., "Influence of industrial sources on children›s health: hot spot studies in North Rhine Westphalia, Germany," International Journal of Hygiene and Environmental Health, vol. 210, no. 5, pp. 591–599, 2007.

22. A. C. Câra, F. Buntinx, M. van den Akker, G.-J. Dinant, and C. Manolovici, "Industrial air pollution and children›s respiratory health: a natural experiment in Călărasi," European Journal of General Practice, vol. 13, no. 3, pp. 135–143, 2007.

23. A. C. Cra, J. Degryse, M. van den Akker, G.-J. Dinant, C. Manolovici, and F. Buntinx, "Impact of early childhood air pollution on respiratory status of school children," European Journal of General Practice, vol. 16, no. 3, pp. 133–138, 2010.

24. World Health, Air Quality Guidelines. Global Update 2005. Particulate Matter Ozone, Nitrogen Dioxide and Sulfur Dioxide, World Health Organization, Copenhagen, Denmark, 2006,http://www.euro.who.int/__data/assets/pdf_file/0005/78638/E90038.pdf.

25. M. De Santis, R. Pasetto, G. Minelli, and S. Conti, "Methods for mortality analysis in SENTIERI Project,"Epidemiologia and prevenzione, vol. 35, supplement 4, no. 5-6, pp. 24–28, 2011 (Italian).

26. R. Pasetto, N. Caranci, and R. Pirastu, "Deprivation indices in small-area studies of environment and health in Italy," Epidemiologia e prevenzione, vol. 35, supplement 4, no. 5-6, pp. 174–180, 2011 (Italian).

27. P. Elliott and D. A. Savitz, "Design issues in small-area studies of environment and health,"Environmental Health Perspectives, vol. 116, no. 8, pp. 1098–1104, 2008.

28. J. Wakefield, "Ecologic studies revisited," Annual Review of Public Health, vol. 29, pp. 75–90, 2008.

29. D. A. Savitz, "Commentary: a niche for ecologic studies in environmental epidemiology," Epidemiology, vol. 23, no. 1, pp. 53–54, 2012.

30. B. Terracini and R. Pirastu, "General guidance to the interpretation of

vital statistics in polluted areas," inHuman Health in Areas with Local Industrial Contamination. Challenges and Perspectives, with Examples From Sicily, WHO Europe, 2013.

31. A. Barchielli, R. De Angelis, and L. Frova, "Mortality data for the study of digestive cancers in Italy: characteristics and quality," Annali dell›Istituto Superiore di Sanita, vol. 32, no. 4, pp. 433–442, 1996 (Italian).

32. C. Bruno, P. Comba, P. Maiozzi, and T. Vetrugno, "Accuracy of death certification of pleural mesothelioma in Italy," European Journal of Epidemiology, vol. 12, no. 4, pp. 421–423, 1996.

33. S. Conti, M. Masocco, V. Toccaceli et al., "Mortality from human transmissible spongiform encephalopathies: a record linkage study," Neuroepidemiology, vol. 24, no. 4, pp. 214–220, 2005.

34. P. Comba, F. Bianchi, S. Conti et al., "SENTIERI Project: discussion and conclusions," Epidemiologia e prevenzione, vol. 35, no. 5-6, pp. 163–171, 2011 (Italian).

35. A. Cernigliaro, A. Marras, S. Pollina Addario, and S. Scondotto, Stato di Salute Della Popolazione Residente Nelle Aree a Rischio Ambientale e nei siti di Interesse Nazionale per le Bonifiche in Sicilia. Analisi dei dati ReNCam (anni 2004–2011) e dei Ricoveri Ospedalieri (anni 2007–2011), Supplemento Monografico Notiziario Osservatorio Epidemiologico, 2013.

36. R. Pasetto, L. Sampaolo, and R. Pirastu, "Measures of material and social circumstances to adjust for deprivation in small-area studies of environment and health: review and perspectives," Annali dell›Istituto Superiore di Sanita, vol. 46, no. 2, pp. 185–197, 2010.

37. D. Mirabelli, D. Cavone, E. Merler et al., "Non-occupational exposure to asbestos and malignant mesothelioma in the Italian National Registry of Mesotheliomas," Occupational and Environmental Medicine, vol. 67, no. 11, pp. 792–794, 2010.

38. J. A. C. Sterne and G. D. Smith, "Sifting the evidence: whats wrong with significance tests?" British Medical Journal, vol. 322, no. 7280, pp. 226–231, 2001.

39. OMS, "Classificazioni delle malattie, traumatismi e cause di morte. Nona revisione," 1975, ISTAT, vol. 1-2, Metodi e Norme, serie C no. 10, 1997.

40. OMS, "Classificazione statistica internazionale delle malattie e dei problemi sanitari correlati," in Decima Revisione, vol. 1–3, Ministero della Sanità, 2001.

41. R. N. Anderson, A. M. Miniño, D. L. Hoyert, and H. M. Rosenberg,

"Comparability of cause of death between ICD-9 and ICD-10: preliminary estimates," National Vital Statistics Reports, vol. 49, no. 2, pp. 1–32, 2001.

42. R. N. Anderson and H. M. Rosenberg, "Disease classification: measuring the effect of the Tenth Revision of the International Classification of Diseases on cause-of-death data in the United States," Statistics in Medicine, vol. 22, no. 9, pp. 1551–1570, 2003.

43. Eurostat, "Guidelines for bridge coding studies. European Statistics on Causes of Death—COD Methodological information. Document prepared by the Task Force on 'ICD-10 up-dates and ACME implementation in Europe'," 2007.

44. "Analisi del Bridge Coding ICD-9 ICD-10 per le statistiche di mortalità per causa in Italia," ISTAT, Metodi e norme, n. 50, 2011.

45. R. Capocaccia and G. Caselli, Popolazione Residente Per età e Sesso Nelle Province Italiane: Anni 1972–1981, Università degli Studi di Roma, Dipartimento di Scienze Demografiche, Roma, Italy, 1990.

46. ISTAT-ISS, Ricostruzione Della Popolazione Residente Per età e Sesso Nelle Province Italiane: Anni 1982–1991, ISTAT, Roma, Italy, 1996.

47. M. De Maria, G. Minelli, and S. Conti, "L'utilizzo dei dati di mortalità a livello comunale in Italia: progetto SENTIERI," Epidemiologia and Prevenzione, vol. 35, supplement 4, no. 5-6, pp. 181–184, 2011.

Chapter 4

THE MEXICAN ENVIRONMENTAL FLOW STANDARD: SCOPE, APPLICATION AND IMPLEMENTATION

María Antonieta Gómez-Balandra, María del Pilar Saldaña-Fabela, Maricela Martínez-Jiménez

Hydrobiology and Environmental Assessment Department, Mexican Institute of Water Technology, Jiutepec, México

ABSTRACT

With the implementation of the Official Mexican Standard NOM-011-CONAGUA-2000 [1], the water balance of 730 basins has been calculated and its water availability agreement is published. This rule points out to allocate water for the environment only as an annual volume since methods for estimating environmental flows were not standardized in the country. For this reason, The Water Agency (CONAGUA) issued the standard NMX-AA- 159-SCFI-2012 [2], to assess environmental flows needed both, at the strategic level in Integrated Water Resources Management (IWRM), or as part of the Environmental Impact Assessment (EIA) of large hydraulic projects. For over ten years, this standard was developed and finally published in September 2012 [3]. It explains different methods from hydrological to holistic approaches, with examples for the country. Its application will cover the urgent need to preserve water for ecosystems in watersheds with high ecological importance and low stress for water use. In this paper, an analysis of the environmental flow standard and examples of the suggested hydrological methods are presented. For its implementation, some steps are taking place, mainly establishing environmental water reserves and building capacities. In addition, environmental allocations are becoming a common practice for all water projects, as well as setting limits to hydrological alterations by hydroelectric dams. The standard promotes the use of technical integration tools to analyze the responses of ecosystems to changes in the flow regime and adaptive management under different scenarios of water use. Although the main steps have been taken, its implementation as mandatory rule will take time.

INTRODUCTION

Environmental, ecological or in stream flow (EF) is the amount of water that is kept flowing down in order to maintain the river in a desired environmental condition in terms of its spatial and timing distribution required to maintain the components, functions and aquatic ecosystems processes from which people obtain benefits ([4-10]. The concept has evolved to recognize the natural variability of flow regimes which consider: magnitude, duration, timing, frequency and rate of change associated to environmental services and ecosystems resilience [11].

Environmental flow assessment (EFA) is based on a scientific process to deal with basin features, probabilistic variability of natural flows and cross section hydromorphology models. These issues determine habitat connectivity and availability to be used by species at several stages of their life cycle. EFA is also a social process since water regulators and users define the level of stress that can be supported by a river in terms of their quantity, quality and ecosystem services, although not many of these decisions have been properly taken. Then, generalized flow standards can be used to allocate water for the environment [12] but, every river can have an environmental flow regime according to what people want from a river in terms of conservation and use [13].

EFA as an escalated set of methodologies is becoming an important toolbox to assess sequentially from rivers' natural flow regime to its ecological integrity (structure, function, processes and alterations). Methods have been classified as: hydrological, hydraulics, for habitat simulation and holistic [9,14-17]). Hydrological approaches are the basis to identify the natural flow regime and its alterations caused by water subtractions or hydraulic infrastructure [11,18]. Results are complemented with hydraulic measurements and associated to habitat availability and connectivity among aquatic, riparian and coastal communities. In addition, holistic approaches consider environmental services for people and comparison of water allocation strategies or altered scenarios by infrastructure [19]. Due to the condition of many rivers which are over-allocated, fragmented, polluted and losing biodiversity, it is recognized that EFA is needed as a part of approaches both at strategic level in the integrated water resource management IWRM and as a part of EIA for large hydraulic projects [20-22]

The EF implementation is on progress and overcoming several drags, since hydrological records are sometimes the only available data for many regions, whereas hydraulic studies are really scarce and its development is time and cost consuming [9]. Besides, these methods need to broaden their scope towards habitat availability for species, not only as hydraulics research. This approach is under development since several disciplines are involved and need to formulate flow alteration—habitat availability—ecological responses hypothesis to be tested through different indicators. The final goal is to deal with an adaptive approach both to operate hydraulic infrastructure and produce conditions for preserving species, maybe in the year by year basis. Thus the final output is to reduce the effects of anthropogenic activities on aquatic ecosystems. Changes can be followed by hydrological parameters, habitat availability, species presence—absence, condition factor, weight—length relationship and food webs, among others [23]

EFA MEXICAN APPROACH

The "environmental water use or use for ecological conservation" is defined by the National Water Law as "the minimum flow or volume required in receiving water bodies, including streams or reservoirs, or the minimum flow of natural discharge from one aquifer that must be maintained to protect environmental conditions and the ecological balance.

The specific definition of Ecological Flow in the Mexican Standard [2], which establishes the procedure to determine the environmental flow is: the quantity, quality, and flow variations or water levels required to preserve environmental services, components, functions, processes and the resilience of aquatic and terrestrial ecosystems. They depend on the hydrological, geomorphological, ecological and social processes. This implies that in addition to provide water for domestic, urban public, livestock and agriculture uses, it is possible to maintain flows from both runoff and aquifer outcropto preserverivers (perennial, intermittent and ephemeral), lenticwater bodies (lakes, ponds, and wetlands), and riparian ecosystems.

The need for ecological flows was a pendant issue since the Water Act publication in 1992 and re -emerged with the application of an Official National Standard [1], to set basins' water availability. This availability comprises a committed natural discharge as a fraction of the hydrogeological unit of natural discharge, committed as superficial water to diverse uses or to be preserved to prevent negative impacts on ecosystems or the migration of bad quality water into the aquifer. The estimated water availability is then published in each hydrometric gauge of the basin and as a total volume.

The Mexican Standard to estimate the EF is a technical guideline rather than a compulsory limit [2]. Its main scientific principles are: the recognition of natural hydrological regime and the gradient of the biological condition.

Therefore, any methodology is valid for the standard as long as it focuses on understanding the ecological significance of each component of the natural flow regime, and it generates proposals for its conservation or restoration in whole or partially, from the functional point of view.

Following a top-down initiative and based on the gradients of ecological importance and water scarcity, an environmental objective (EO) was assigned such as water availability is issued for each basin or sub-basin in the country (Figure 1). The Rio Verde sub-basins were numbered for further analysis.

The EF Mexican standard calls for the application of hydrological methods as the basis to reserve or allocate water for the environment. Therefore, according to the EO's, reference values (% of medium annual volume or flow) as well as the components of a seasonal flow regime are stated as follows:

1) Under a modified Tennant approach (García et al. [5]).
a) Percentage of mean annual flow (MAF) with different seasonal percentages (Reference values).
b) Percentage of monthly mean flows MMF.
c) A base flow during dry season not less than the historical minimum monthly flow (MinF).
2) Under WWF Mexico approach.
a) A total volume associated to an ordinary regime (TVOR), considering wet, medium and dry years.
b) A total volume related to the frequency of floods' regime (TVFR) with a return period of 1, 1.5 and 5 years to be reproduced within 10 years return period according to EO's.
3) Under TNC approach [14].
a) Intra and inter-annual natural variability.
b) Thresholds for hydrological alterations on hydrological parameters with ecological importance (monthly flows, 1, 7, 30 and 90 days minimum and maximum flows).

These approaches can be applied at basins, sub-basins or river reaches as it is pointed out in Figure 2. At least 20 years of hydrological data are recommended for all these analysis.

The Mexican flow standard also describes some initial experiences in using habitat simulation with BBM building blocks methodology [24]. as holistic approach and recommends the use of habitat simulation methods like PHabSim or others.

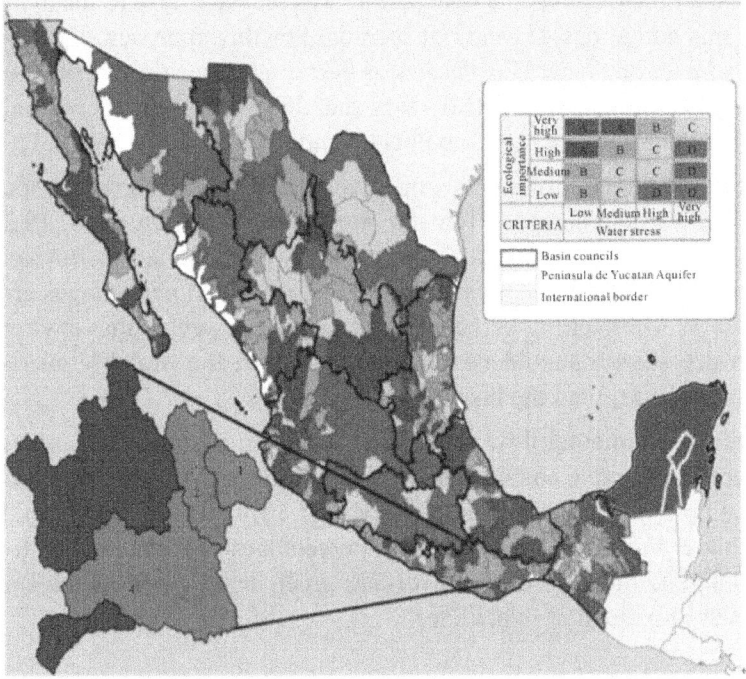

Figure 1: Environmental objectives for the Mexican basins and the Rio Verde Basin (Modified from WWF, [24]).

EFA Hydrological Methods Application

The Environmental objectives for the Rio Verde Basin and sub-basins located in the hydrological region 20 named Costa Chica de Guerrero were stated as is pointed out in Table 1.

The hydrological gauge Paso de la Reina was chosen to show result from the standard application (Figure 3) in this site a proposed hydro project is under planning studies.

Modified Tennant Approach ([5])

General steps:

1) Selection of the site of study.
2) Analysis of the monthly dataset.
3) Determination of monthly and annual flow regime.
4) Formulation of proposals for monthly and annual environmental flow regime.

As mentioned above, the selected site was the Paso de la Reina hydrometric gauge, this gauge has 45 years of records. For this approach, the start of the monthly series begins in May because is the month when the rain season begins. The monthly flow regime (MMF) was stated as well as the mean annual flow (MAF) as the reference started points (Figure 4).

According to environmental objectives (EO), Table 2 shows recommended seasonal flow percentages. Tennant modified by García et al. [5,26], and proposed by [25]. As Paso de la Reina site was classified as Environmental Objective "A" [2], then annual and monthly allocated percentages are shown in Figure 5. After setting percentages it must be reviewed that environmental flow in dry season should never be greater than the monthly average flow (MMF) nor lesser than the base flow.

The environmental flow proposal is based on monthly percentages (light green area in Figure 6 and values of Table 3). For dry years the proposal can be adjusted up to seasonal thresholds (30% and 60% of MAF blueline). For rain season these limits can be set between percentages of mean annual flow MAF (blue line) and monthly mean flow (dark green area). For each stream, a base flow must be estimated (blackline).

WWF Approach

According to the World Wildlife Fund Mexico approach a reference volume as percentage of the medium annual flow should be equal or greater than 40% (Table 4).

The next step is to identify alterations in the hydrological regime trough dividing time series data in two periods. For the first set of 20 years (natural hydrological regimen RHN) obtain percentiles 10 and 90 and check if the number of the monthly means of the second period (present hydrological regime RHA) that are within these limits. If the fulfillment of the actual regime (RHA) <50% is in relation to percentiles (RHN) the river is considered altered. The Figures 7 and 8 pointed out that the Rio Verde at Paso de la Reina hydrometric gauge is not altered.

For seasonal environmental flow estimation, the Total Ordinary Volume Regime (TOVR) and the Total Volume for Flood Regime (TVFR) were obtained using the procedure proposed by World Wildlife Fund Mexico at the study basin level. TOVR was obtained multiplying the frequency by volume for each year condition to be reproduced with regards to the environmental objective (Table 5).

The two steps to calculate TVFR that try to resemble natural pattern, according to the conservation objectives, were:

1) Classification and characterization of peaks: To separate flood types initially used the criterion of magnitude, maximum flow and probability of occurrence over time (frequency). Analysis of statistical distributions are needed to determine magnitude of floods associated return periods as follows:

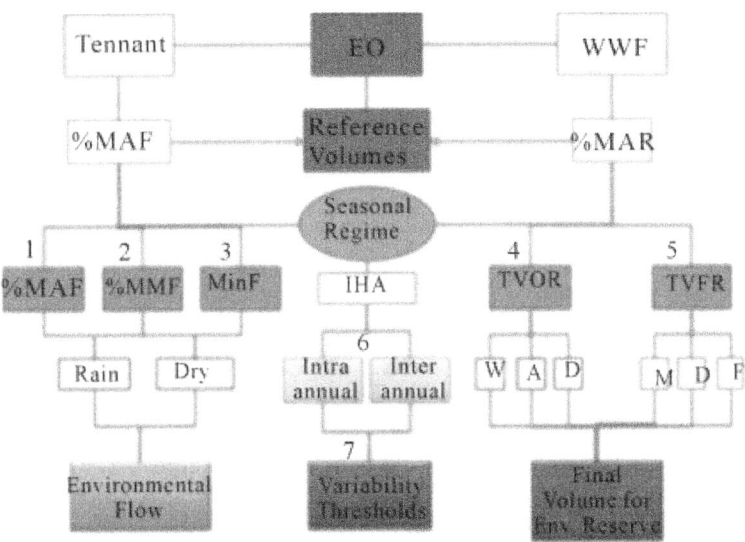

Figure 2: EFA hydrological approaches in the Mexican standard.

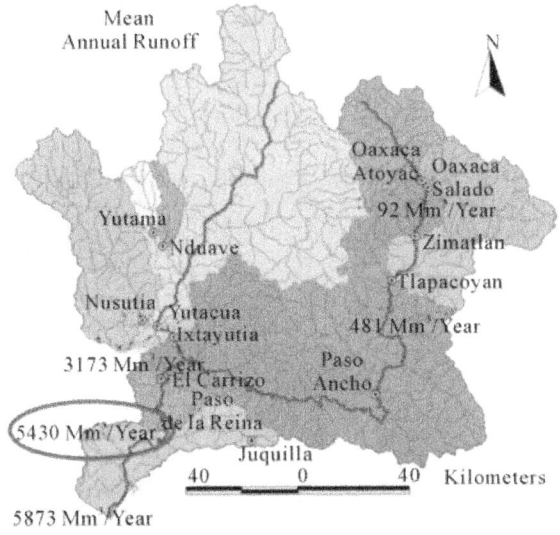

Figure 3: Water availability issued for the basin. CFR: 05/ 07/2013 [25].

Table 1: Environmental objectives by sub-basin

Rio Verde Sub-basins	1	2	3	4	5
Annual Availability (Mm³/year)	92.4	481.1	3173.8	5430.3	5873.3
Ecological Importance	High	High	High	Medium	High
Water use stress	Medium	Medium	Low	Low	Low
Present/Desired Condition	Good	Good	Very Good	Good	Very Good
Environmental Objective	B	B	A	B	A

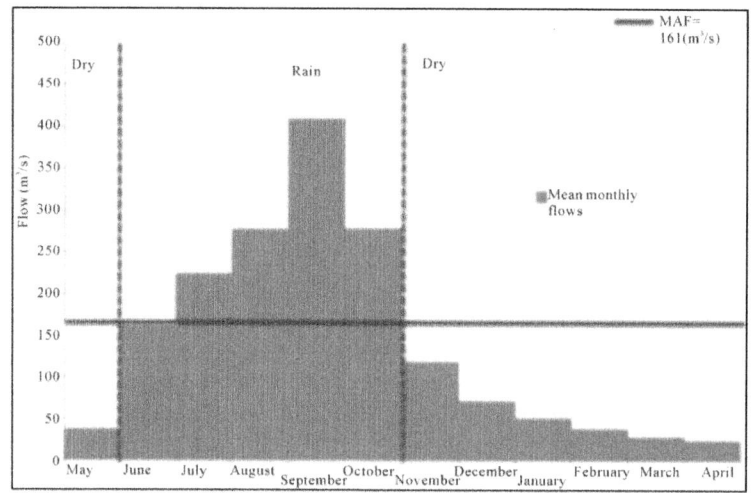

Figure 4: Monthly flow regime and mean annual flow that determines the dry and rainy seasons.

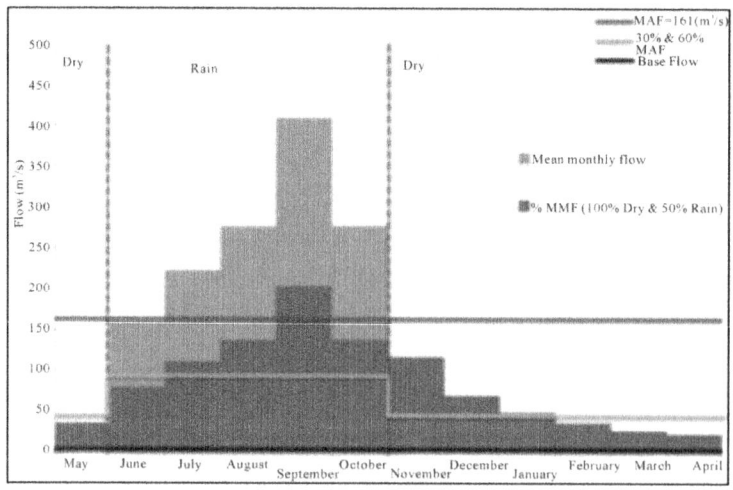

Figure 5: Annual and monthly allocated percentages.

a) Runoff with return period of 1 year (Category I).
b) Runoff with return period of 1.5 years (Category II).
c) Runoff with return period of 5 years (Category III).

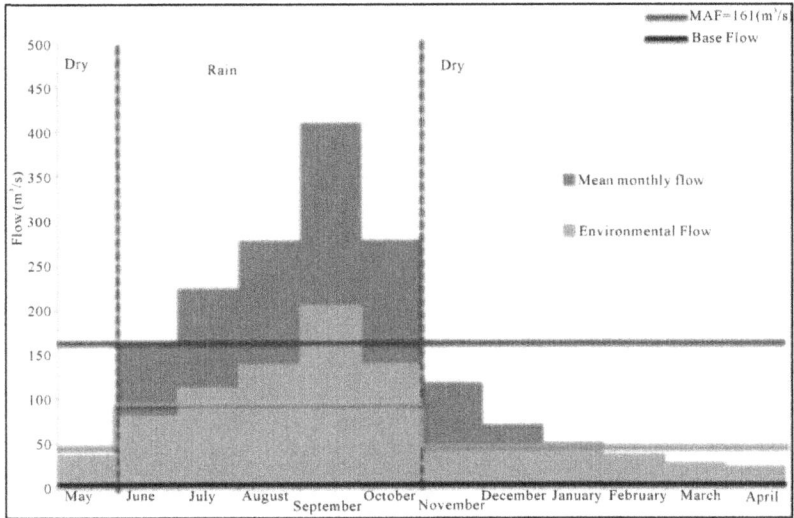

Figure 6: Environmental flow proposal.

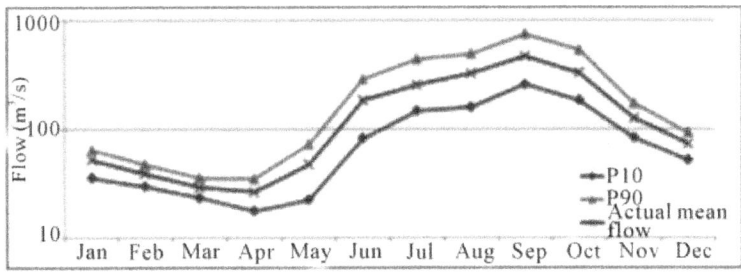

Figure 7: Monthly present mean and RHN limits.

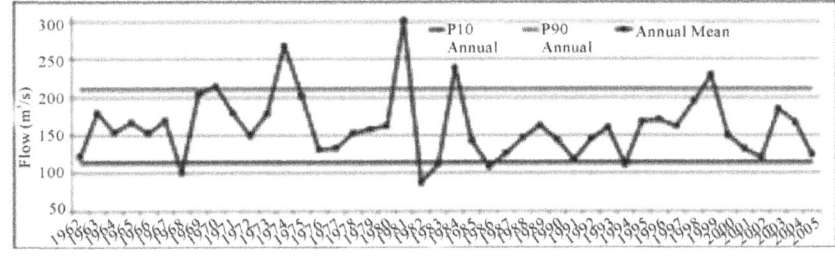

Figure 8: Annual present mean and RHN limits.

2) Duration in days of each runoff period is then obtained and the product of this duration, frequency and volume resulted in the Total Volume for Flood Regime for three different runoffs within 10 years. Table 6 shows the results for these calculations.

The integration of the final reserved volume (FRV) or environmental flow to be considered in the annual water balance of the basin and its water availability is the sum of total ordinary volume regime (TOVR) with the total volume of flood regime (TVFR), as follows:

Table 2: Reference values for each EO

Environmental Objective	Period			
	Dry		Rain	
	% MAF	% MMF	% MAF	% MMF
A	30	100	60	50
B	20	80	40	40
C	15	60	30	30
D	5	40	10	20

Table 3: Monthly environmental flow regime in cubic meters per second (m³/s)

Month	Environmental Flow
January	48.3
February	37.8
March	28.3
April	24.2
May	37
June	80.4
July	111.9
August	138.7
September	204.8
October	139.2
November	48.3
December	48.3

Table 4: Reference values for specific environmental objectives

Environmental Objective	Conservation Gradient	Environmental Flow (% MAF) Perennial streams
A	Very good	>40 > 64.6 m³/s or >2172 Mm³/year

Table 5: Total Ordinary Volume Regime

Type of Year	Very dry	Dry	Average	Wet
Percentile	P0	P10	P25	P75
Ordinary volume regime (OVR Mm³/year)	2052	2862	3611	6066
% Mean Annual Runoff	40	56	71	119
Frequency	0.2	0.3	0.4	0.1
Total Ordinary Volume Regime	3319 Mm³/year			

FRV = TOVR + TVFR
FRV = 3319 + 198 = 3517 Mm³/year

Table 6: Total Volume of Flood Regime

Attribute of hydrological regime		Category I	Category II	Category III
Magnitude		1 year return	1.5 years return	5 years return
	m³/s	303	843	1874
	Mm³/day	26	73	162
Frequency		10	6	2
Duration		3	2	1
Timing		Jul-Oct		
Rate of change	Rise	75		
	Fall	40		
Total Volume for Flood Regime in 10 Years		1980 Mm³		
TVFR/Year		198 Mm³		

MAR = 5430 Mm³/year
FRV = 3517 Mm³/Year = 64% of MAR

CONCLUSIONS

Although the Mexican Water Act states the environmental use and the Availability Standard points out an annual volume to the environment in each basin, for more than fifteen years, the minimum flow as much as 10% of monthly flows have prevailed to deal with environmental allocation of water for projects with consumptive uses (water supply and irrigation). Availability agreements have been issued and are going to be reviewed every two years. Therefore, implementing EFA is the next challenge to be included in these agreements and then gives some priority of reserving water for the environment [27].

The Mexican environmental flow standard as a technical procedure gives the support to move ahead from this 10% minimum approach and consider the natural flow variability. Through the recognition of this natural variability and applying at least hydrological methods, twofold protection goals can be achieved to: 1) Reproduce seasonal and even monthly variability and 2) Consider thresholds for alteration either seasonal, monthly or inter-annually when ecological process can be vulnerable.

The hydrological methods are also the basis for the regional water resources management under frameworks like the integrated watershed resource management IWRM and the ecological limits of hydrological alteration (ELOHA) [28]

The application of habitat simulation and holistic methods is under development in Mexico. Therefore, experiences at regional and basin level as well as for large infrastructure projects are an urgent need to protect the ecological integrity of ecosystems. These approaches should be integrative and under an interdisciplinary approach.

As a result of the application of habitat simulation or holistic methods, a set of ecological and social indicators can be obtained to follow during the EF implementation. For Mexico, the regional ecosystem changes and social responses will be valuable when the adaptive approach has to consider tradeoffs between ecological conservation and water use tightening further hydraulic development.

REFERENCES

1. NOM-011-CNA-2000, "Norma Oficial Mexicana, Conservación del Recurso Agua. Que Establece las Especificaciones y el Método para Disponibilidad Media Anual de las Aguas Nacionales," DOF 17-Abril-2002, México, p. 20.
2. NMX-AA-159-SCFI-2012, "Norma Mexicana. Que Establece el

Procedimiento para la Determinación del Caudal Ecológico en Cuencas Hidrológicas," DOF 20 de Septiembre de 2012, México, p. 123.

3. P. A. E. Lis, M. A. G. Balandra and P. S. Fabela, "Requerimientos para Implementar el Caudal Ambiental en México," Instituto Mexicano de Tecnología del Agua, Alianza World Wildlife Fund/Fundación Gonzalo Río Arronte. Programa Hidrológico Internacional UNESCO, SEMARNAT, 2007, p. 176.

4. T. Annear, I. Chisholm, H. Beecher and A. Locke, "Instream flows for Riverine Resource Steweardship," Revised Edition, Instream Flow Council, Cheyenne, 2004, p. 268.

5. E. García, R. González, P. Martínez, J. Athala and G. Paz-Soldán, "Guía de Aplicación de los Métodos de cálculo de Caudales de Reserva Ecológicos en México, Colección," CNA-IMTA-SEMARNAP, México D.F., 1999, p. 190.

6. S. Postel and B. Richter, "Rivers for Life: Managing Water for People and Nature," Island Press, Washington DC, 2003.

7. N. L. Poff, "Managing for Variability to Sustain Freshwater Ecosystems," Journal of Water Resources Planning and Management, Vol. 135, No. 1, 2009, pp. 1-4.

8. A. H. Arthington, S. E. Bunn, N. L. Poff and R. J. Naiman, "The Challenge of Providing Environmental Flow Rules to Sustain River Ecosystems," Ecological Applications: A Publication of the Ecological Society of America, Vol. 16, No. 4, 2006, pp. 1311-1138.

9. R. E. Tharme, "A Global Perspective on Environmental Flow Assessment: Emerging Trends in the Development and Application of Environmental Flow Methodologies for Rivers," River Research and Applications, Vol. 19, No. 5-6, 2003, pp. 397-441. http://dx.doi.org/10.1002/rra.736

10. M. Acreman and M. J. Dunbar, "Defining Environmental River Flow Requirements a Review," Hydrology and Earth System Sciences, Vol. 8, No. 5, 2004, pp. 861-876. http://dx.doi.org/10.5194/hess-8-861-2004

11. B. D. Richter, J. V. Baumgartner, R. Wigington and D. P. Braun, "How Much Water Does A River Need?" Freshwater Biology, Vol. 37, No. 1, 1997, pp. 231-249. http://dx.doi.org/10.1046/j.1365-2427.1997.00153.x

12. B. D. Richter, "Short Communication Re-Thinking Environmental Flows: From Allocations and Reserves to Sustainability Boundaries," River Research and Applications, Vol. 26, No. 8, 2010, pp. 1052-1063.

13. J. O'Keeffe and T. Le Quesne, "Keeping Rivers Alive— A Primer on Environmental Flows and Their Assessment," World Wildlife Found—

Water Security Series 2, Surrey, 2009.
14. TNC (The Nature Conservancy), "Indicators of Hydrologic Alteration," Version 7.1 User's Manual, 2009, p. 81.
15. WWF and Fundación Gonzalo Río Arronte, "Guía Para la Determinación de Caudal Ecológico en México Sistematización de Experiencias de la Alianza WWF," Fundación Gonzalo Río Arronte I.A.P., Mexico D.F., 2011.
16. M. J. DUNBAR and M. C. Acreman, "Applied Hydro-Ecological Science for the Twenty-First Century," In: M. C. Acreman, Ed., Hydro-Ecology: Linking Hydrology and Aquatic Ecology (Proceedings of an International Workshop (HW2). Centre for Ecology and Hydrology (Formerly Institute of Hydrology), Wallingford, Oxfordshire, 2001, pp. 1-18.
17. R. Hirji and R. Davis, "Environmental Flows in Water Resources Policies, Plans and Projects. Case Studies," Water Resources, No. 117. Washington DC, 2009, p. 181.
18. N. L. Poff, J. D. Allan, M. A. Palmer, D. D. Hart, B. D. Richter, A. H. Arthington, K. H. Rogers, J. L. Meyer and J. A. Stanford, "River Flows and Water Wars: Emerging Science for Environmental Decision Making," Frontiers in Ecology and the Environment, Vol. 1, No. 6, 2003, pp. 298-306. http://dx.doi.org/10.1890/1540-9295(2003)0010298:RFAWWE.2.0.CO;2
19. A. H. Arthington, R. J. Naiman, M. E. McClain and C. Nilsson, "Preserving the Biodiversity and Ecological Services of Rivers: New Challenges and Research Opportunities," Freshwater Biology, Vol. 55, No. 1, 2009, pp. 1-16. http://dx.doi.org/10.1111/j.1365-2427.2009.02340.x
20. C. Nilsson and B. M. Renöfält, "Linking Flow Regime and Water Quality in Rivers: A Challenge to Adaptive Catchment Management," Ecology and Society, Vol. 13, No. 2, 2008, 20 p.
21. C. J. Vörösmarty, P. B. McIntyre, M. O. Gessner, D. Dudgeon, A. Prusevich, P. Green, S. Glidden, S. E. Bunn, C. A. Sullivan, C. R. Liermann and P. M. Davies, "Global Threats to Human Water Security and River Biodiversity," Nature, Vol. 467, 2010, pp. 555-561. http://dx.doi.org/10.1038/nature09440
22. The Brisbane Declaration, The Brisbane Declaration at 10th International Riversymposium & Environmental Flows Conference, Brisbane, 2007.
23. H. Chen, L. Ma, W. Guo, Y. Yang, T. Guo and C. Feng, "Linking Water Quality and Quantity in Environmental Flow Assessment in Deteriorated Ecosystems: A Food Web View," Plos One, Vol. 8, 2013, No. 7, p. 13.

24. WWF, Fundación Río Arronte, and CONAGUA, "Norma Mexicana de Caudal Ecológico—Una PolíTica Pública Para la Gestión del Agua a Través de la Conservación del Régimen Hidrológico," Mexico D.F., 2012.
25. CONAGUA, "Identificación de Reservas Potenciales de agua Para el Medio Ambiente en México," Comisión Nacional del Agua, Alianza WWF-Fundación Gonzalo Río Arronte I.A.P, Primera Ed. Mexico, 2011, p. 87.
26. D. L. Tennant, "Instream Flow Regimens for Fish, Wildlife, Recreation and Related Environmental Resources," Fisheries, Vol. 1, No. 4, 1976, pp. 6-10. http://dx.doi.org/10.1577/1548-8446(1976)0012.0.CO;2
27. J. King, R. Tharme and M. de Villier, "Environmental Flow Assessments for Rivers: Manual for the Building Block Methodology," Water Research Commission, SAF, Cape Town, 2008.
28. N. L. Poff, B. D. Richter, A. H. Arthington, S. E. Bunn, R. J. Naiman, E. Kendy, M. Acreman, C. Apse, B. P. Bledsoe, M. C. Freeman, J. Henriksen, R. B. Jacobson, J. G. Kennen, D. M. Merritt, J. H. O'Keeffe, J. D. Olden, K. Rogers, R. E. Tharme and A. Warner, "The Ecological Limits of Hydrologic Alteration (ELOHA): A New Framework for Developing Regional Environmental Flow Standards," Freshwater Biology, Vol. 55, No. 1, 2010, pp. 147-170. http://dx.doi.org/10.1111/j.1365-2427.2009.02204.x

Chapter 5

EVALUATION OF SOIL HYDRAULIC PARAMETERS IN SOILS AND LAND USE CHANGE

Fereshte Haghighi[1], Mirmasoud Kheirkhah[2] and Bahram Saghafian[2]

[1]Soil Conservation and Watershed Management Institute, Tehran
[2]Soil Conservation and Watershed Management Research Institute, Tehran, Iran

INTRODUCTION

The knowledge of soil water properties and land-use effects on these properties are important for efficient soil and water management. Furthermore, the use of the pedo transfer functions (PTFs) to estimate soil water content (θ_h) is important to assess. The loosening effect of dryland farming on soil water retention is known. In this chapter we review soil water content, pedotransfer functions and some infiltration models applicability for two land-use types. The land-use effect on soil water retention may be significant at water potentials of -33 kPa and 0 kPa in the soil. At the -1500 kPa pressure head, water content may not be affected by cultivation of rangeland at different soil depths. In addition, pedotransfer functions can be used as a physically based model for soil water retention characterization in the various areas. Moreover, it is essential to evaluate the infiltration models applicability for different soils and various land-uses.

Definition

Soil hydraulic properties

Soil hydraulic properties govern transport processes and water balance in soils. Water retention capacity, infiltration rate, and saturated hydraulic conductivity are important soil hydraulic properties. Soil water retention and saturated hydraulic conductivity (K_s) are necessary input data for the simulations of water flow in soil and water engineering. Characterizing hydrological behavior of catchments requires knowledge of hydraulic parameters.

Soil Water Retention

Soil water retention at field capacity (FC) and permanent wilting point (PWP) are used to estimate the water depth applied by irrigation (Hansen et al., 1980), and to calculate water availability, as a crucial factor to assess the land area suitability for crop producing (Sys et al., 1991).

SOIL WATER RETENTION CAPACITY AND LAND USE

One important soil hydraulic property is water retention capacity, which affects soil productivity and management. Soil water content (θ_h) governs the transport characteristics of water and solutes in soils. The knowledge of water retention capacity and land use effects on this property is important for efficient soil and water management. Upon conversion of natural lands to cultivated fields, water retention capacity is strongly influenced (Schwartz et al., 2000; Bormann and Klaassen, 2008; Zhou et al., 2008). Soil water retention at field capacity (FC) and permanent wilting point (PWP) are important to estimate the irrigation water depth which may be affected by land use change. Soil water retention characteristic, is affected by soil organic matter (SOM) content and porosity, which are significantly influenced by land use type (Zhou et al., 2008).

We conducted a study to evaluate, document, and quantify the effect of cultivation of rangeland on soil water retention in field capacity (FC), permanent wilting point (PWP), and to test the use of the van Genuchten equation to estimate θ_h in cultivated and natural lands in the same soils of the Taleghan watershed in Iran.

Significant differences in the OM and bulk density (BD) were observed between dryland farming and rangeland at both depths of 0 cm - 15 cm and 15 cm - 30 cm. Soil sample water contents at different pressure heads under both land use types are presented in Figures 1. The overall measured and fitted soil water retention curves did not show significant difference within the selected water potentials for both land use types in this study. However, measured θ_s (0 kPa) values were found to be significantly lower for dryland farming when compared with rangeland at depths of 0 cm - 15 cm and 15 cm - 30 cm, respectively. Moreover, the land use effect on soil water retention was significant at a water potential of -33 kPa (FC) based on laboratory measurements only at the top (15 cm depth). The results indicated that the conversion of rangeland to dryland farming led to a significant decrease (16.56% on average) in the FC at a depth of 0 cm - 15 cm. The mean -1500 kPa (PWP) water content was not affected by the land-use type. Figure 1 indicates that the mean total field capacity (FC) was significantly greater in rangeland when compared with dryland farming at a depth of 0 cm - 15 cm. In this study, there were not statistically significant differences in water content at other potentials (-50 kPa, -100 kPa, -500 kPa, and -1000 kPa pressures) between the two types of land use presented in Figure 1. At those pressure heads and at a -1500 kPa water content, the amount of micropores were not affected by cultivation of rangelands (Fig. 1). Overall, the results showed that the soil pore system and reduced total porosity under dryland farming can decrease water storage capacity at water potentials of -33 kPa and 0 kPa. Ndiaye et al., (2007) has shown that improper soil management decreases the soil macroporosity in the long-term affecting the θ_s. The data obtained in our study demonstrated the loosening effect of dryland farming on soil water retention. Previous studies on the effect of land use have demonstrated clear changes in soil physical properties, such as soil porosity, SOM, and BD, in relation to hydraulic properties (Bormann and Klassen 2008; Haghighi et al., 2010b).

PEDOTRANSFER FUNCTIONS (PTFS)

Determination of soil water properties required as input data for simulation models is time consuming and relatively costly (Wösten et al., 1995). Thus, indirect estimation of these characteristics has been proposed as one alternative to direct estimation of the soil hydraulic parameters based on the measured water retention data. Pedotransfer functions (PTFs) are emerged as the relationship between soil hydraulic and other more available measured properties (Bouma, 1989) which can be used to estimate hydraulic parameters. PTFs are useful tools for modeling applications.

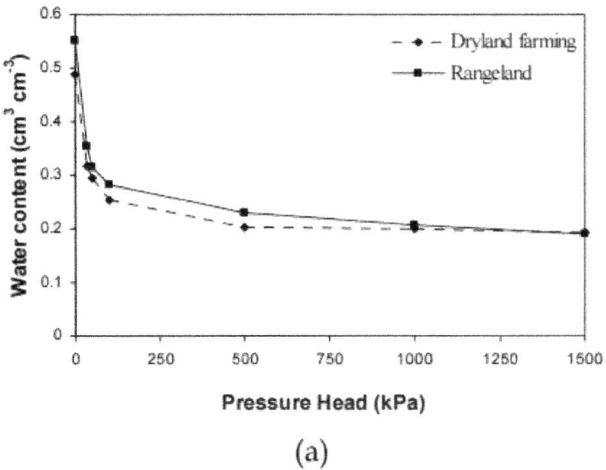

(a)

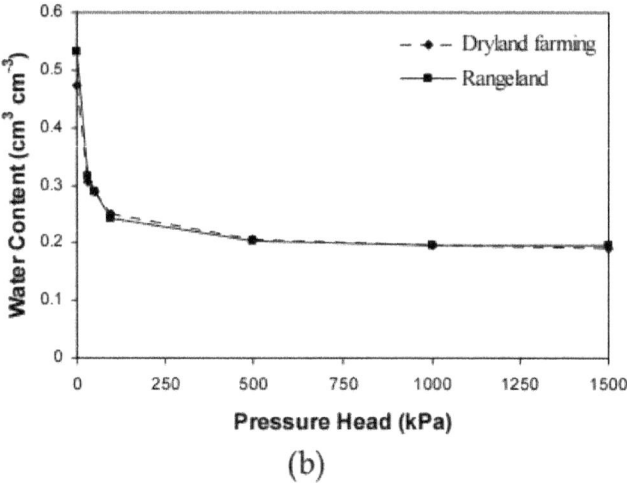

(b)

Figure 1: Soil water content as a function of the pressure head for two landuse types at depths of (a) 0 cm -15 cm and (b) 15 cm - 30 cm.

PEDOTRANSFER FUNCTIONS (PTFS) AND DIFFERENT LAND USES

To estimate the land use effects on soil water retention, van Genuchten model (Van Genuchten, 1980) may be applied. Some researches have correlated van Genuchten parameters with soil organic matter, bulk density (BD), and soil particle size distribution and many researchers have estimated the water retention curve using soil texture, bulk density, and porosity.

Many statistical equations (pedotransfer functions) characterizing the water retention curve have been presented (Kutilek and Nielsen., 1994). PTFs are useful tools for modeling applications. Such analytical functions are derived involving various soil data. Such data are measured in the field and laboratory analysis. Soil hydraulic parameters derived through PTFs can be used to express soil hydraulic properties and water retention (Brooks and Corey, 1964). Consequently, physically based models such as van Genuchten representing a pedotransfer function may be considered as a valuable tool to simulate the soil water properties in different land uses.

The $\theta_{(h)}$ data may be fitted to van Genuchten equation to derive retention curves and parameters (a, n, and θ_r), using the RETC (RETention curve) optimization computer code (Van Genuchten et al., 1991). The van Genuchten model (van Genuchten, 1980) is defined as:

$$\theta(h) = \theta_r + \frac{\theta_s - \theta_r}{(1+|\alpha h|^n)^m} \quad \theta_{(h)} = \theta_s \quad h \geq 0 \tag{1}$$

Where $\theta_{(h)}$ (cm³ cm⁻³) is the volumetric water content (for h<0), θ_r (cm³ cm⁻³) is the residual water content, and θ_s (cm³ cm⁻³) is the saturated water content. Here, m is $1-(1/n)$ with $n>1$. α (cm⁻¹) and n are empirical parameters determining the shape of the curve which were obtained for each core. Parameter n is related to steepness of the water retention curve.

$$K(S_e) = K_s S_e [1 - (1 - S_e^{(1/m)})^m] \tag{2}$$

Where K_s (mm/h) is saturated hydraulic conductivity and S_e is the effective saturation expressed as:

$$S_e = \frac{\theta - \theta_r}{\theta_s - \theta_r} \tag{3}$$

The effect of landuse type on soil water retention and PTF applications have not been documented for different land-uses to the best of our knowledge. In developing countries, there is a lack of large databases that are needed to develop PTFs. Thus, in many developing countries, the use of available PTFs can cause errors for estimating soil hydraulic properties. This encourages further investigations of the model applications and development of suitable point and parametric PTFs for estimating soil hydraulic properties in the studied area. The selection of more suitable PTFs for application where there are not developed PTFs caused by a lack of large databases is difficult. Consequently, it is essential to evaluate the model applicability and to develop point and parametric PTFs for estimating soil hydraulic properties for the soils in various sites. Thus, the estimates may be improved by comprehensive local studies.

Table 1: Mualem-van Genuchten parameters calculated for old grassland (site A), recently reseeded grassland (site B) and previous maize cultivated land (site C) (Sonneveld et al, 2003)

Location (site)	Depth: 10–20 cm					Depth: 25–50 cm				
	K_{sat} (cm/day)	θ_{sat} (–)	α (1/cm)	l (–)	n (–)	K_{sat} (cm/day)	θ_{sat} (–)	α (1/cm)	l (–)	n (–)
A	16.0	0.43	0.007	0.944	1.37	23.1	0.43	0.014	−0.350	1.41
B	16.0	0.42	0.007	0.331	1.36	15.8	0.40	0.009	−0.448	1.33
C	17.4	0.40	0.008	0.093	1.35	14.7	0.38	0.010	−0.482	1.33

EVALUATION OF COMMON INFILTRATION MODELS FOR DIFFERENT LAND-USES

The evaluation of infiltration characteristics as a hydrologic process in soils is necessary in agricultural studies. The knowledge of final steady infiltration rate is important for irrigation water efficiency, designing desirable irrigation systems, and loss of water. Thus, infiltration rate is important factor in sustainable agriculture, effective watershed management, surface runoff, and retaining water and soil resources. Since measuring the final infiltration rate is time consuming, several physical and empirical models have proposed to determine it. The empirical models such as Kostiakov (1932) and Horton (1940), and physical model such as Philip (1957) are the most common models to estimate infiltration rate of the soils.

Kostiakov-Lewis model

The model of Kostiakov modified for long times as follows:

$$f = at^{-b} + f_c \tag{4}$$

Where a and b are the equation's parameters (a>0 and 0<b<1) i_c is the steady infiltration rate (LT^{-1}).

Horton's Model

The Horton's infiltration model (Horton, 1940) is expressed as follows:

$$f = (f_0 - f_c)e^{-kt} + f_c \tag{5}$$

Where i_c is the presumed final infiltration rate (LT^{-1}), i_0 is the initial infiltration rate (LT^{-1}) and t is time (T). k is the infiltration decay factor.

Philip Two-parameter Model

The Philip two-term model is expressed as (Philip, 1957):

$$f = \frac{1}{2}St^{-0.5} + A \tag{6}$$

Where f is the infiltration rate (LT^{-1}) as a function of time.

A= Transmissivity factor (LT^{-1}) as a function of soil properties and water contents,

S = Sorptivity that is function of soil matric suction ($LT^{-0.5}$).

t= time (T)

Singh (1992) expressed that the various models can estimate different values of the final infiltration rate in a soil which seems to be uncorrect, because of the final infiltration rate is a soil-dependent factor. Compared to the previous investigations on soil infiltration properties and models, studies on soil infiltration modelling depending on land use are scarce. Nevertheless, it can be assumed that landuse type have a significant impact on soil infiltration and infiltration models performance. Machiwal et al (2006) observed the infiltration process was well described by the Philip's model in a wasteland of Kharagpur, India. However, different soil management that influences the final infiltration rate is a major reason for different applicability of these models. Long-term effects of land use changes on soil infiltration and infiltration models (e.g., Horton, Kostiakov, and Philip models) can be observed (Navar and Synnott, 2000; Shukla et al, 2003).

Thus, the variability of soil infiltration characteristics and goodness of fit of the infiltration models for different land-uses should be considered during infiltration modelling studies helping on correct predictions of final infiltration for different land uses. Ability of these models for estimating the infiltration rate in different land-uses and soil management has been examined by some researchers. Gifford (1976) observed among the Horton, Kostiakov and Philip's models, the Horton's model was the best model to fit the infiltration data in mostly semi-arid rangelands from the Australia, but only under specific conditions. Shukla et al (2003) evaluated some of the infiltration models at different soil management and landuse systems in Ohio and observed among infiltration models, the Swartzendruber model was the best ones and fitted the observed infiltration data with lower sum of squares and higher model efficiency. Davidoff and Selim (1986) examined the goodness of fit for eight infiltration models on a Norwood soil with four winter cover crop treatments and results of their study showed that the Philip, Kostiakov and Horton's models had best predictions than the other models. Haghighi et al (2010b) evaluated the effects of rangeland and dryland farming land uses on performance of some infiltration models to estimate the final infiltration rate of soils. The study was conducted on some soils of Taleghan watershed, Iran. According to reports (Taleghan watershed study report, 1993), investigated soils are calcareous and classified as Typic Xerorthents. Mean annual rainfall alters from 464 to 796 mm and lands slope is by 15 %. The soil texture varied from clay-loam to silty clay loam.

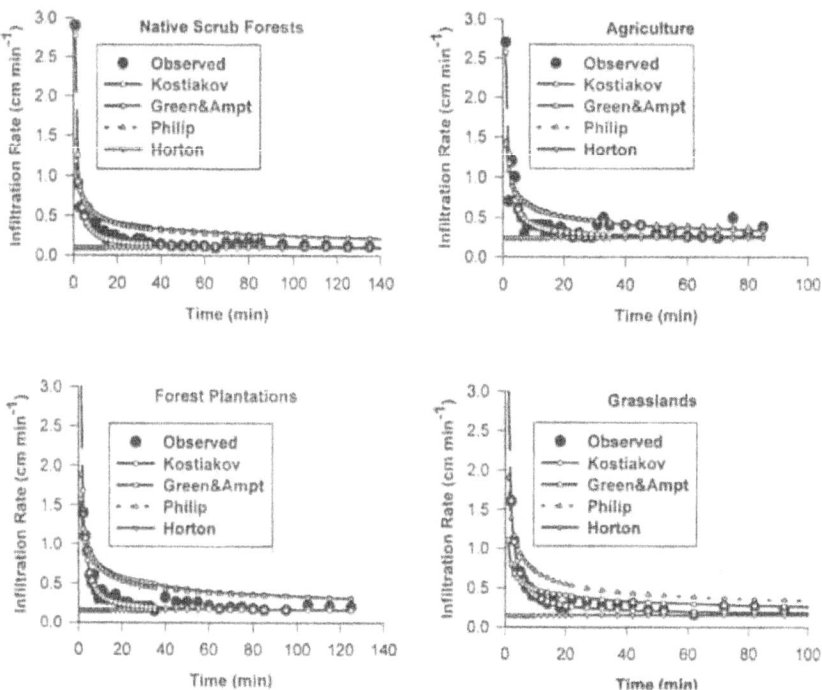

Figure 2: Effects of land use changes on soil infiltration and infiltration models (Navar and Synnott, 2000).

In our study, the goodness of the fit of selected models and ability of them for estimating the final infiltration rate of rangeland and dryland farming soils was evaluated using the root mean squared errors (RMSE). The values of R^2 were determined high (0.99) and equal for all sites and land-uses, but the values of RMSE and the final steady infiltration showed that the estimated infiltration rates by the infiltration model of Horton, approached more closely to the measured ones at the selected area

Table 1.. The Horton's model was the best model selected for both of land-uses. It can be expressed that various models can suppose different final infiltration (f_c) values for a soil, which seems to be not practical, because fc is a soildependent parameter, in general. Common changes in land-use negatively affect soil physical properties and decrease soil infiltration rate and could change modelling performance. Effect of land-use should be well documented aiming on good predictions in the studied areas and elsewhere.

The infiltration models can be used for estimating the infiltration rate in soils, well. But only, one or some of these models are better and appropriate for a specific site. Thus, the infiltration models should be analyzed for their ability to estimate the infiltration rate of each location. The investigation of Haghighi et al (2010a) showed that the Horton's model is the best ones selected for rangeland and dryland farming and land-use type is not an important factor to affect infiltration models efficiency. Due to a few number of investigation in this field of research, there is a need for further investigation on land-use effect on infiltration modelling and for the impact of land-use on soil infiltration characteristics, as well.

Table 2: Parameters of the selected infiltration models in both of land use types (Haghighi et al, 2010a)

Land use	Kostiakov-Lewis model			Philip two-term model		Horton's model			Observed final infiltration rate (cm min^{-1})
	c	b	a	S	A	β	f_0	f_c	
Rangeland 1	0.2052	0.543	1.136	1.217	0.223	0.130	0.894	0.299	0.2803
Rangeland 2	0	0.770	1.058	1.322	0.235	0.021	0.596	0.190	0.2441
Rangeland 3	0.0870	0.606	0.839	0.970	0.127	0.090	0.556	0.183	0.1613
Rangeland 4	5.13×10^{-14}	0.855	0.5076	0.543	0.208	0.045	0.387	0.225	0.2285
Dryland farming 1	0.0585	0.636	1.475	1.741	0.162	0.085	0.909	0.258	0.2347
Dryland farming 2	0	0.522	2.863	2.984	0.018	0.120	1.601	0.200	0.1863
Dryland farming 3	1.37×10^{-12}	0.781	0.4857	0.592	0.119	0.029	0.290	0.120	0.1098
Dryland farming 4	0.0452	0.681	0.2555	0.315	0.073	0.071	0.198	0.089	0.0869

CONCLUSIONS

Soil management and land use change may affect soil water retention at a -33 kPa (FC) potential in the soil based on laboratory measurements and model simulations. Lower water content at the -33 kPa potential would be expected upon conversion of natural lands to cultivated lands. In addition, the saturated soil water content (θ_s) may be affected by cultivation of rangeland. Moreover, because cultivation of natural lands affects soil macroporosity, we suggest measuring soil water retention at higher suction heads to document the land use effect on soil water retention properties in relation to soil macropores. Appropriate technology for dryland farming and suitable measures are necessary to improve soil water retention where cropping is required.

The findings show that the van Genuchten model is useful in describing soil water retention. Thus, use of this model may be considered as a valuable tool to gain more knowledge of hydraulic properties for various soil types. The effect of land use type on soil water retention and PTF applications have not been documented for dryland farming to the best of our knowledge. In many developing countries, such as Iran, the use of available PTFs can cause errors for estimating soil hydraulic properties. This review encourages further investigations of the model applications and development of suitable point and parametric PTFs for estimating soil hydraulic properties. The selection of more suitable PTFs for application where there are not developed PTFs caused by a lack of large databases is difficult. Consequently, it is essential to evaluate the model applicability and to develop point and parametric PTFs for estimating soil hydraulic properties for different land uses.

REFERENCES

1. Bormann, H., Klaassen, K., 2008. "Seasonal and land use dependent variability of soil hydraulic and soil hydrological properties of two Northern German soils." Geoderma. 145, 295-302.
2. Brooks, R.H., Corey, A.T., 1964. "Hydraulic properties of porous media." Civil Engineering Dept., Colorado State University, Fort Collins, CO.
3. Davidoff, B. and Selim, H.M., 1986. Goodness of fit for eight water infiltration models, Soil Sci. Soc. Am. J. 50, 759-764
4. Gifford, GF. 1976. Applicability of some infiltration formulae to rangeland infiltrometer data. Journal of Hydrology. 28, 1-11.
5. Haghighi F, Gorji M, Shorafa M, Sarmadian F, Mohammadi MH. 2010. "Evaluation of some infiltration models and hydraulic parameters." Spanish Journal of Agricultural Research (INIA), 8 (1), 210-217.
6. Haghighi F, Gorji M, Shorafa M. 2010. "A study of the effects of land use change on soil physical properties and organic matter." Land Degradation & Development Journal. In press 10.1002/ldr.999.
7. Horton, R.E. 1940. An approach toward a physical interpretation of infiltration-capacity. Soil Sci. Soc. Am. Proc.5, 339-417.
8. Kostiakov, A. N. 1932. On the dynamics of the coefficient of water percolation in soils and on the necessity of studying it from a dynamic point of view for the purposes of amelioration.
9. Kutílek M. 2004. Soil hydraulic properties as related to soil structure, Soil Till. Re 79, 175–184.
10. Machiwal, D., K.J.., and B.C. Mal. 2006. Modelling infiltration and quantifying spatial soil variability in a watershed of Kharagpur, India. Biosystems Engineering, 95(4), 569-582.
11. Mishra, s.k, J.V.Tyagi, and V.P.Singh. 2003. Comparison of infiltration models. Hydrological processes. 17, 2629-2652 .
12. Navar.J, and T.J. Synnote, 2000. Soil infiltration and land use in Linares, N.L., Mexico, Terra Latinoamericana. 18 (3), 255-262.
13. Ndiaye, B., Molenat, J., Hallaire, V., Hamon, C.G.Y., 2007. "Effects of agricultural practices on hydraulic properties and water movement in

soils in Brittany (France)". J. Soil & Tillage Res. 93, 251-263.
14. Philip, J.R. 1957. The theory of infiltration: 4. Sorptivity and algebric infiltration equations. Soil Sci. 84, 257-264.
15. Schwartz, R.C., Unger, P.W., Evett S.R., 2000. "Land use effects on soil hydraulic properties." ISTRO.
16. Shukla, M.K., R.Lal., and P.Unkefer. 2003. Experimental evaluation of infiltration models for different land use and soil management systems. Soil Sci. 168 (3), 178-191.
17. Singh V.P. 1992. Elementary Hydrology. Prentice Hall: Englewood Cliffs, NJ.
18. Sonneveld MPW., Bachx MAHM., Bouma.J. 2003. Simulation of soil water regimes including pedotransfer functions and land use related preferential flow. Geoderma, 112, p: 97-110.
19. University of Tehran. "Taleghan Watershed study report"., 1993. Irrigation Engineering Department, Iran.
20. Van Genuchten, M.Th., 1980. "A closed-form equation for predicting the hydraulic conductivity of unsaturated soils." Soil Sci. Soc. Am. J. 44, 892–898.
21. Van Genuchten, M.Th., Leij, F. J., Yates, S.R., 1991. "The RETC code for quantifying the hydraulic functions of unsaturated soils." EPA/600/2-91/065. R. S. Kerr, Environmental Research Laboratory. U. S. Environmental Protection Agency, Ada, OK. p: 83.
22. Wösten, J.H.M., Finke, P.A., Jansen, M.J.W., 1995. "Comparison of class and continuous pedotransfer functions to generate soil hydraulic characteristics." Geoderma. 66, 227–237.
23. Zhou, X., Lin, H.S., White, E.A., 2008. "Surface soil hydraulic properties in four soil series under different land uses and their temporal changes." Catena. 73, 180-188.

Chapter 6

ENERGY-RELATED CARBON EMISSIONS OF CHINA'S MODEL ENVIRONMENTAL CITIES

Kevin Lo
Department of Resource Management and Geography, University of Melbourne, Melbourne, VIC 3010, Australia

ABSTRACT

This paper identifies three types of model environmental cities in China and examines their levels of energy-related carbon emissions using a bottom-up accounting system. Model environmental cities are identified as those that have been recently awarded official recognition from the central government for their efforts in environmental protection. The findings show that, on average, the Low-Carbon Cities have lower annual carbon emissions, carbon intensities, and per capita emissions than the Eco-Garden Cities and the Environmental Protection Cities. Compared internationally, the Eco-Garden Cities and the Environmental Protection Cities have per capita emissions that are similar to those of American cities whereas per capita emissions from the Low-Carbon Cities are similar to those of European cities. The result indicates that addressing climate change is not a priority for some model environmental cities. Policy changes are needed to prioritize climate mitigation in these cities, considering that climate change is a cross-cutting environmental issue with wide-ranging impact.

INTRODUCTION

Climate change is one of the most significant challenges facing the world today and cities are increasingly seen as key contributors to the issue. Urban activities, such as motorized transport, industrial production, electricity generation, domestic fuel use, and waste disposal generate significant amounts of greenhouse gases [1–3]. The International Energy Agency (IEA) forecasted that, by 2030, urban areas will account for 76% of global carbon emissions [4].

Thus, to prevent dangerous climate change, cities must bear some responsibility for reducing their impact on climate change and formulating effective responses [1, 5]. Inventorying urban carbon emissions is an important first step to address climate change effectively and fairly. Creating emissions inventories at the city level allows policymakers to identify the sources, establish baselines, monitor changes over time, make cross-comparisons with other localities, set appropriate emissions reduction targets, and formulate appropriate solutions [1]. The importance of inventorying urban carbon emissions is underscored by the large number of studies dedicated to such task [6–12].

China is critically important to the global effort of addressing climate change because of its enormous size of emissions [13]. Since 2006, the government has stepped up measures to promote energy conservation and renewable energy [14, 15]. Yet virtually no cities in China openly publish their greenhouse gases inventory on a regular basis. A number of studies have attempted to fill this gap. The calculation by Li et al. [16] showed that Shanghai's energy-related carbon emissions increased from 110 million tCO_2e in 1995 to 180 million tCO_2e in 2006. Wang et al. [17], also studying Shanghai, found that the city's total carbon emissions during 2000–2008 increased from 136 teragrams of CO_2e to 200 teragrams of CO_2e. In 2008, per capita emissions in Shanghai reached 14.03 tCO_2e, which were higher than both the world average and the average in China. Bi et al. [18] calculated that, in 2009, the total carbon emissions in Nanjing reached 75.43 million tCO_2e and per capital emissions reached 9.78 tCO_2e. Inventorying 12 Chinese cities, Wang et al. [19] found that carbon emissions in all cities rose from 2004 to 2008, but per capita emissions varied widely in different cities. Although China had lower-than-global-average per capita emissions (5.5 tCO_2e), the per capita emissions of many Chinese cities were greater than 8.0 tCO_2e, which were comparable to or even higher than those of many other cities in developed countries. Dhakal [20] approximated the energy-related emissions from 35 major cities using provincial carbon intensity data. The results showed that there were large differences in terms of per capita emissions within these cities, with the carbon-intensive cities largely located in the central and western parts of China, an area that attracted energy-intensive industries with low energy prices and favourable policies [21].

Building on these previous findings, this study aims to examine the levels of energy-related carbon emissions from China's model environmental cities, identified as those that have been recently awarded official recognition from the central government for their efforts in environmental protection. These cities are regarded as the leaders of environmental protection and sustainable development. However, concrete evidence on the environmental qualities of these cities in the present era of climate change has been lacking. As such, one may wonder whether and to what extent these cities are really environmentally friendly from the perspective of low-carbon urbanism. Following this introduction, this paper identifies three types of modelenvironmental cities. It then estimates the carbon emissions from these cities using annual emissions, annual per capita emissions, and carbon intensity. The paper concludes with a discussion of the policy implications of the results.

THREE TYPES OF MODEL ENVIRONMENTAL CITIES

This paper defines model environmental cities as cities that have received official recognition from the central government for their efforts in environmental protection. Although many cities in China declare their environmental credentials, cities that received official recognition have to go through quality assurance system and meet certain specified environmental standards. Therefore, they are considered as the leaders in China with respect to environmental issues. There are three types of model environmental cities: Eco-Garden Cities, Environmental Protection Cities, and Low-Carbon Cities. The remaining portion of this section will examine these different types of model environmental cities more closely.

Eco-Garden Cities

The Eco-Garden City Program was established in 2004 by the Ministry of Construction (now the Ministry of Housing and Urban-Rural Development [MOHURD]). The aim of the program is to award recognition to cities that protect and improve the health of urban ecosystems. Eco-Garden Cities are required to work toward a number of objectives, including construction and maintenance of parkland, improving access for pedestrian and bicyclist, urban greening, improving public transport, increasing the density of developed areas, protection of ecosystems, increasing the proportion of energy-efficient buildings, and increasing the deployment of renewable energy. The MOHURD developed a complicated system of 90 quantitative targets to guide local governments toward the achievement of these objectives. For example, one of the energy and climate targets is to achieve at least 60% coverage by public transportation. Because not every target is compulsory, an Eco-Garden City may fail to meet one or more of the criteria.

A city aspiring to be an Eco-Garden City must first establish an action plan to achieve the specified targets and then must implement the plan for at least three years. The city is then evaluated, first by the Department of Housing and Urban-Rural Development at the provincial-level and then by the MOHURD. The assessment process involves independent surveys, remote-sensing analysis, field inspections, and expert evaluations. A city that passes the assessment earns the right to call itself an Eco-Garden City. However, an Eco-Garden City that is later found to be noncompliant with the requirements would be either cautioned or disqualified, depending on the degree, scale, and nature of the infringement.

Environmental Protection Cities

In 1997, the Environmental Protection Bureau (now the Ministry of Environmental Protection [MEP]) established the Environmental Protection City Program to encourage cities to improve environmental quality through sustainable development. Environmental Protection Cities are required to work toward a number of environmental objectives, including air pollution prevention, water pollution prevention, improving waste management, reducing industrial energy intensity, and increasing the deployment of renewable energy use. The MEP developed 22 quantitative targets for Environmental Protection Cities. All of the targets are compulsory, unlike those of the Eco-Garden City Program. There are two targets relevant to climate change. First, industrial energy intensity must be less than the national average. Second, more than 50% of the energy used must come from sources other than coal.

A city aspiring to be an Environmental Protection City must first prepare a proposal demonstrating concrete plans to meet the specified targets and then submit the proposal to the provincial Environmental Protection Bureau for approval. The city must then implement the proposal and later submit a detailed report on how the criteria have been met. The MEP then assembles a team of experts to evaluate the application. The application is immediately rejected if more than two targets have not been met. However, if only one or two targets are not met, the city is given three months to meet the targets. A program revision in 2011 introduced a five-year lifecycle for the program, under which the Environmental Protection City status automatically expires after five years unless the city reapplies successfully.

Low-Carbon Cities

The National Development and Reform Commission (NDRC) established the Low-Carbon City Pilot Program in 2010 as a key climate change response. The NDRC is a powerful ministry in the central government. It is responsible for planning the economic development of the country and has been designated as the ministry with primary responsibility for energy and climate change. Participating local governments are encouraged to go beyond mere compliance with existing low-carbon policies and programs. The specific obligations and commitments include developing a long-term low-carbon development plan, implementing institutional reform and effective policy instruments to lower carbon emissions, developing low-carbon industries, buildings and transportation, developing carbon emissions accounting and management systems, and promoting low-carbon lifestyles and consumption. The NDRC has yet to release quantified targets for these obligations. The NDRC has promised to provide venues for information exchange and policy learning to help cities achieve their objectives.

Studied Cities

At the time of writing, there were 12 Eco-Garden Cities, among them, 9 were prefecture-level cities (Guilin, Yangzhou, Nanjing, Suzhou, Qingdao, Weihai, Jincheng, Hangzhou, and Shaoxing), and 3 were county-level cities. There were 22 Environmental Protection Model cities; among them 14 were prefecture-level cities (Dongguan, Foshan, Zhongshan, Langfang, Daqing, Yichang, Zhenjiang, Xuzhou, Huaian, Yinchuan, Liaocheng, Linyi, Weihai, and Shaoxing). There were 8 cities in the Low-Carbon Cities Pilot Program and 6 of them were prefecture-level cities (Xiamen, Shenzhen, Guiyang, Baoding, Nanchang, and Hangzhou) and 2 were provincial-level cities (Chongqing and Tianjin). Because of overlaps and the lack of accessibility to statistical yearbooks in some cities, in total, there are 15 cities that serve as the dataset for this study. Figure 1 displays the locations of the studied cities.

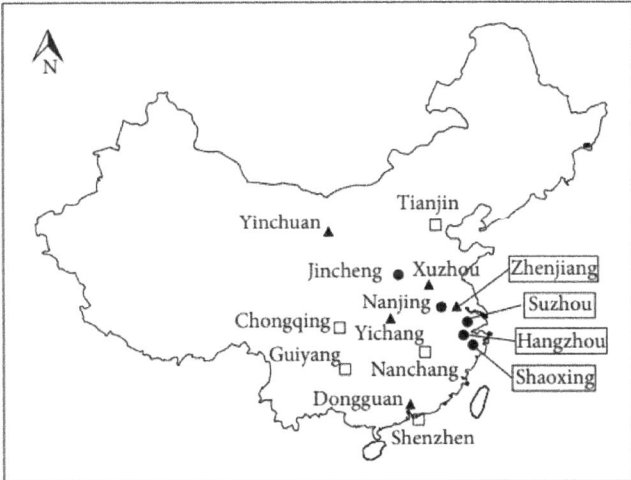

☐ Low-Carbon Cities
▲ Environmental Protection Cities
● Eco-Garden Cities

Figure 1: The locations of the studied cities.

METHODS

The present study uses a bottom-up accounting system to measure carbon emissions from stationary and mobile sources at the municipal level. It first calculates stationary emissions from the combustion of primary energy sources (e.g., coal and gasoline). The inventory is then modified because a city may import or export electricity. Emissions from a net export of electricity are deducted from the inventory, and emissions associated with a net import of electricity are added to the inventory. The formula for stationary emissions is

$$\text{GHG} = \left(\sum_{i,j} E_{i,j} C_j \right) + (I - X) C_e, \tag{1}$$

where GHG is the total stationary emissions in tons; i represents subsectors (e.g., textile or steel making); represents the energy type (e.g., coal or crude oil); $E_{i,j}$ is the energy consumption per subsector and energy type; C_j is the CO_2 emission factor for specific energy types; I is the quantity of electricity imported into the city; X is the quantity of exported electricity; and C_e is the CO_2 emission factor for electricity.

With regard to mobile emissions, this study only calculates emissions from road transportation because of a lack of data from air, water, and rail transportation. Air, water, and rail transportation is also usually cross-boundary, and it is therefore difficult to attribute their emissions to a particular city. This study distinguishes five modes of road transport: taxis, public buses, other passenger vehicles, trucks, and motorcycles. Each mode of transport has its own unique annual mileage and emission factors. Emissions from road transport are calculated using the following equation:

$$\text{GHG} = \sum_i P_i \times D_i \times C_g, \tag{2}$$

where GHG is carbon emissions from road transportation; i is the mode of transport (e.g., taxis or passenger cars); P_i is the number of vehicles in each mode of transportation; D_i is the annual mileage of a type- I vehicle measured in km; and C_g is the CO_2 emission factor measured in ton/km.

Data describing industrial energy consumption; import and export of electricity; numbers of taxis, buses, passenger vehicles, trucks, and motorcycles; GDP; and population are collected from the 2011 statistical yearbook for each municipality [22–36]. Some of the statistical yearbooks do not contain information about export and import of electricity, in which case the import/export data are estimated from the provincial-level electricity balance table that is in the China Energy Statistical Yearbook [37]. Carbon emission factors are obtain from the 2006 IPCC Guidelines for National Greenhouse

Gas Inventories (Table 1) [38]. Carbon emission factors for electricity are acquired from Wang et al. [39], who calculated the emission factors in China's six large national power grids using electricity generation fuel mixes and power exchange data (Table2). Average annual mileage and average gasoline consumption for different modes of transportation are collected from Loo and Li [3] (Table 3).

Table 1: Emission factors for combustion of fuel

Fuel	CO_2 emission factor (kg/TJ)	Fuel	CO_2 emission factor (kg/TJ)
Crude oil	73,300	Natural gas	56,100
Gasoline	74,100	Other petroleum products	73,300
Kerosene	71,500	Raw coal	97,500
Fuel oil	77,400	Washed coal	94,600
Diesel oil	74,100	Other types of washed coal	94,600
Coke oven gas	44,400	Coke	107,000
LPG	63,100		

Table 2: Emission factors of six major power grids

Grid	CO_2 emission factor (ton/million kWh)	Grid	CO_2 emission factor (ton/million kWh)
Northeast China	871	East China	688
North China	874	Northwest China	701
Central China	555	South China	552

Table 3: Average annual mileage and average gasoline consumption different modes of transportation

	Taxis	City buses	Private cars and institutional vehicles	Motorcycles
Average annual mileage (km)	71,175	34,000	18,000	10,000
Average gasoline consumption (km/kg)	15.07	3.91	15.07	50.23

This study focuses on energy-related emissions because the data are mostly complete and reliable. Three types of emissions are not calculated in this study, including emissions from agriculture, forest management, and other land uses (AFOLU), such as CO_2 from deforestation CH_4 from enteric fermentation and manure, and N_2O from fertilizer use, landfill gases, mainly CH_4, and emissions from industrial processes, such as CO_2 and perfluorocarbon emissions from aluminum smelting and N_2O emissions from the production of nitric acid. These omissions result in an underestimation of the emissions inventories, although not by a significant amount, because energy consumption is the dominant source of carbon emissions in Chinese cities, typically responsible for over 90% of the total emissions [40].

RESULTS

Annual carbon emissions, annual per capita carbon emissions, and carbon intensity of the studied cities are calculated using the methods and data described above. Per capita emissions and carbon intensity are used to facilitate comparisons among the cities, which vary widely in population size and state of economic development. Per capita emissions are calculated by dividing annual emissions by the total population of the city in question. Carbon intensity expresses the emissions generated in the production of 10,000 RMB worth of GDP and is calculated by dividing annual emissions by annual GDP. The results are shown in Figures 2, 3, and 4.

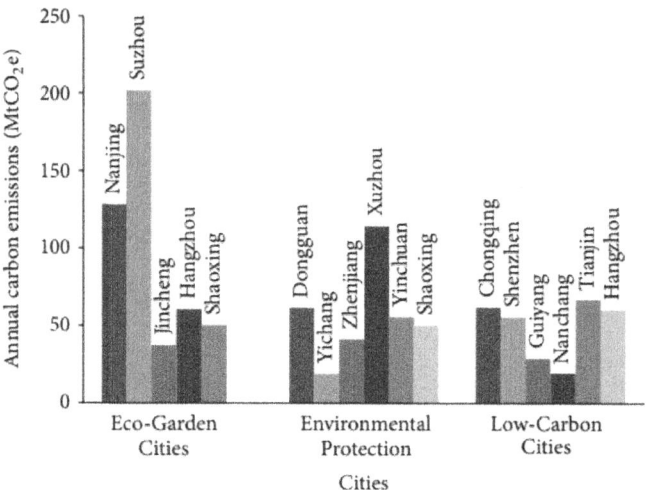

Figure 2: Annual carbon emissions from the model environmental cities.

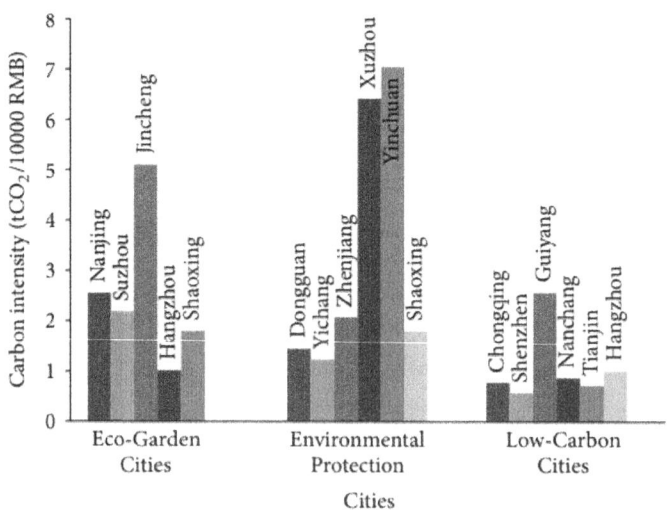

Figure 3: Carbon intensity of the model environmental cities.

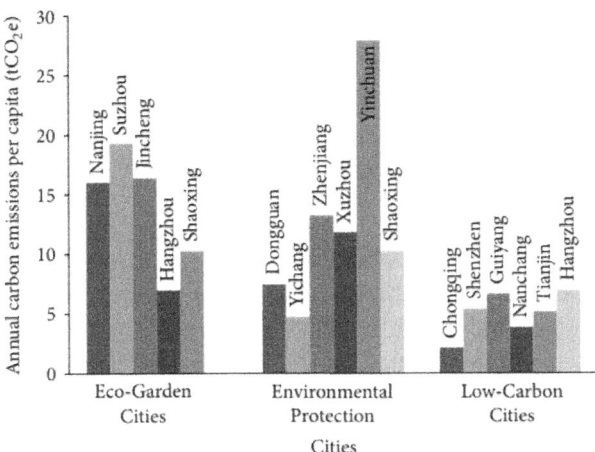

Figure 4: Annual carbon emissions per capita from the model environmental cities.

On average, the Eco-Garden Cities emit far more CO_2 (mean = 95,771,862 tCO_2e) than both the Environmental Protection Cities (mean = 57,102,423 tCO_2e) and the Low-Carbon Cities (mean = 48,991,821 tCO_2e). However, there are significant differences within each group according to the size and economic activity of the city. Nanjing (an Eco-Garden City), Suzhou (an Eco-Garden City), and Xuzhou (an Environmental Protection City), which produce more emissions than any other city in this study, are all located in Jiangsu, one of China's most heavily industrialized area. Yichang, an Environmental Protection City, and Nancheng, a Low-Carbon City, have the lowest levels of emissions. Both of these cities are small-sized cities in inland China.

The Environmental Protection Cities have the highest average carbon intensity (mean = 3.34 $tCO_2/10000\,RMB$), followed by the Eco-Garden Cities (mean = 2.54 $tCO_2/10000\,RMB$) and the Low-Carbon Cities (mean = 1.09 $tCO_2/10000\,RMB$). The most carbon-intensive city is Yinchuan, a coal-dominated Environmental Protection City in the northwest. The least carbon-intensive city is Shenzhen, a wealthy and postindustrial Low-Carbon City in the Pearl River Delta area.

The Low-Carbon Cities have on average substantially lower per capita emissions (mean = 5.03 tCO_2e) than both the Eco-Garden Cities (mean = 13.79 tCO_2e) and the Environmental Protection Cities (mean = 12.58 tCO_2e). Yinchuan is the city with the highest per capita emissions, and, Chongqing, a Low-Carbon City in inland China, has the lowest. Compared internationally, the Eco-Garden Cities and the Environmental Protection Cities have per capita emissions that are similar to American cities, such as Boston (12.23 tCO_2e, 2009 data) and Seattle (11.47 tCO_2e, 2008 data) [41] whereas the levels of per capita emissions from the Low-Carbon Cities are similar to those of European cities, such as Berlin (5.86 tCO_2e, 2007 data) and London (5.98 tCO_2e, 2008 data) [41].

DISCUSSION AND CONCLUSION

The analysis has shown that, on average, the Low-Carbon Cities have lower annual carbon emissions, carbon intensities, and per capita emissions than the Eco-Garden Cities and the Environmental Protection Cities. This reflects the different goals and priorities of the model environmental cities, which are influenced by the programs that award official designations. The Low-Carbon City Program focuses on climate change, and all participating cities are required to implement a number of climate mitigation measures. All Low-Carbon Cities have therefore formulated climate change action plans and committed themselves to carbon intensity targets. Some cities, such as Hangzhou and Nanchang, have announced targets that are more ambitious than the national commitment of reducing carbon intensity by 40–45% of 2005 levels by 2020. Low-Carbon City climate change action plans are usually comprehensive, covering emissions from key sectors, including industry, energy, housing, and transportation. For example, Shenzhen will develop new industries in alternative energy, information technology, biotechnology, new materials, cultural industry, and energy conservation services, promote the deployment of natural gas, nuclear power, and solar energy, and introduce compulsory solar water heating for buildings less than 12 stories in height [42]. Shenzhen is currently the most successful city in the promotion of electric vehicles. The city launched the world's largest green taxi fleet of 300 electric-only vehicles in 2010 and plans to build 250 charging stations and install 12,500 charging posts in the city.

In contrast, climate change is not a focus of the Eco-Garden City Program and the Environmental Protection Program. The Eco-Garden City Program focuses primarily on protecting the health of urban ecosystems. The Environmental Protection City Program concentrates on air and water pollution and waste management. Both programs incorporate the concept of sustainable development and therefore emphasize a broad range of issues, such as economic and social development and environmental quality. While these two programs have energy and climate objectives, such objectives are too narrow and not sufficiently demanding. Furthermore, the objectives can be ignored completely in the Eco-Garden City Program because they are not compulsory. Therefore, the Eco-Garden Cities and the Environmental

Protection Cities have, in general, failed to pay sufficient attention to climate mitigation.

Climate change is a cross-cutting issue that impacts a wide range of economic, social, and environmental aspects of cities. Climate change exacerbates urban heat island effects [43, 44], increases the frequency of flooding due to rising sea levels, leads to more frequent and severe extreme weather events [45], and exacerbates water security problems by changing the amount, variability, timing, form, and intensity of precipitation [46]. For these reasons, climate change mitigation should be made a priority for all types of model environmental cities in China, not just the Low-Carbon Cities. This can be achieved by reforming the Eco-Garden City Program and the Environmental Protection City Program to incorporate more climate-related objectives. These climate objectives should be comprehensive, covering emissions from the industry, building, and transportation sectors, and be sufficiently challenging to motivate cities to experiment with different climate mitigation policies.

REFERENCES

1. H. Bulkeley, Cities and Climate Change, Routledge, Oxon, UK, 2013.
2. K. Lo, "Energy conservation in China's higher education institutions," Energy Policy, vol. 56, pp. 703–710, 2013.
3. B. P. Y. Loo and L. Li, "Carbon dioxide emissions from passenger transport in China since 1949: implications for developing sustainable transport," Energy Policy, vol. 50, pp. 464–476, 2012.
4. International Energy Agency, World Energy Outlook 2008, International Energy Agency, Paris, France, 2008.
5. H. Bulkeley and V. C. Broto, "Government by experiment? Global cities and the governing of climate change," Transactions of the Institute of British Geographers, vol. 38, no. 3, pp. 361–375, 2012.
6. L. Parshall, K. Gurney, S. A. Hammer, D. Mendoza, Y. Zhou, and S. Geethakumar, "Modeling energy consumption and CO_2 emissions at the urban scale: methodological challenges and insights from the United States," Energy Policy, vol. 38, no. 9, pp. 4765–4782, 2010.E. L. Glaeser and M. E. Kahn, "The greenness of cities: carbon dioxide emissions and urban development," Journal of Urban Economics, vol. 67, no. 3, pp. 404–418, 2010.C. Kennedy, J. Steinberger, B. Gasson et al., "Methodology for inventorying greenhouse gas emissions from global cities," Energy Policy, vol. 38, no. 9, pp. 4828–4837, 2010.T. Hillman and A. Ramaswami, "Greenhouse gas emission footprints and energy use benchmarks for eight US cities," Environmental Science & Technology,

vol. 44, no. 6, pp. 1902–1910, 2010.J. Minx, G. Baiocchi, T. Wiedmann et al., "Carbon footprints of cities and other human settlements in the UK," Environmental Research Letters, vol. 8, no. 3, Article ID 035039, 2013.

7. A. Chavez and A. Ramaswami, "Progress toward low carbon cities: approaches for transboundary GHG emissions› footprinting," Carbon Management, vol. 2, no. 4, pp. 471–482, 2011.C. Kennedy, J. Steinberger, B. Gasson et al., "Greenhouse gas emissions from global cities," Environmental Science & Technology, vol. 43, no. 19, pp. 7297–7302, 2009. ·

8. J. S. Gregg, R. J. Andres, and G. Marland, "China: emissions pattern of the world leader in CO_2 emissions from fossil fuel consumption and cement production," Geophysical Research Letters, vol. 35, no. 8, Article ID L08806, 2008.K. Lo and M. Y. Wang, "Energy conservation in China›s Twelfth Five-Year Plan period: continuation or paradigm shift?" Renewable and Sustainable Energy Reviews, vol. 18, pp. 499–507, 2013.

9. K. Lo, "A critical review of China›s rapidly developing renewable energy and energy efficiency policies,"Renewable and Sustainable Energy Reviews, vol. 29, pp. 508–516, 2014.

10. L. Li, C. Chen, S. Xie et al., "Energy demand and carbon emissions under different development scenarios for Shanghai, China," Energy Policy, vol. 38, no. 9, pp. 4797–4807, 2010.

11. Y. Wang, W. Ma, W. Tu, Q. Zhao, and Q. Yu, "A study on carbon emissions in Shanghai 2000–2008, China," Environmental Science & Policy, vol. 27, pp. 151–161, 2013.

12. J. Bi, R. Zhang, H. Wang, M. Liu, and Y. Wu, "The benchmarks of carbon emissions and policy implications for China›s cities: case of Nanjing," Energy Policy, vol. 39, no. 9, pp. 4785–4794, 2011.H. Wang, R. Zhang, M. Liu, and J. Bi, "The carbon emissions of Chinese cities," Atmospheric Chemistry and Physics, vol. 12, pp. 6197–6206, 2012.

13. S. Dhakal, "Urban energy use and carbon emissions from cities in China and policy implications," Energy Policy, vol. 37, no. 11, pp. 4208–4219, 2009.K. Lo, "Deliberating on the energy cap in China: the key to a low-carbon future?" Carbon Management, vol. 4, no. 4, pp. 365–367, 2013.

14. Nanjing Municipal Statistics Bureau, Nanjing Statistical Yearbook 2011, China Statistics Press, Beijing, China, 2011.

15. Suzhou Municipal Statistics Bureau, Suzhou Statistical Yearbook 2011,

China Statistics Press, Beijing, China, 2011.

16. Jincheng Municipal Statistics Bureau, Jincheng Statistical Yearbook 2011, China Statistics Press, Beijing, China, 2011.

17. Hangzhou Municipal Statistics Bureau, Hangzhou Statistical Yearbook 2011, China Statistics Press, Beijing, China, 2011.

18. Shaoxing Municipal Statistics Bureau, Shaoxing Statistical Yearbook 2011, China Statistics Press, Beijing, China, 2011.

19. Dongguan Municipal Statistics Bureau, Dongguan Statistical Yearbook 2011, China Statistics Press, Beijing, China, 2011.

20. Yichang Municipal Statistics Bureau, Yichang Statistical Yearbook 2011, China Statistics Press, Beijing, China, 2011.

21. Zhenjiang Municipal Statistics Bureau, Zhenjiang Statistical Yearbook 2011, China Statistics Press, Beijing, China, 2011.

22. Xuzhou Municipal Statistics Bureau, Xuzhou Statistical Yearbook 2011, China Statistics Press, Beijing, China, 2011.

23. Yinchuan Municipal Statistics Bureau, Yinchuan Statistical Yearbook 2011, China Statistics Press, Beijing, China, 2011.

24. Chongqing Municipal Statistics Bureau, Chongqing Statistical Yearbook 2011, China Statistics Press, Beijing, China, 2011.

25. Shenzhen Municipal Statistics Bureau, Shenzhen Statistical Yearbook 2011, China Statistics Press, Beijing, China, 2011.

26. Guiyang Municipal Statistics Bureau, Guiyang Statistical Yearbook 2011, China Statistics Press, Beijing, China, 2011.

27. Nanchang Municipal Statistics Bureau, Nanchang Statistical Yearbook 2011, China Statistics Press, Beijing, China, 2011.

28. Tianjin Municipal Statistics Bureau, Tianjin Statistical Yearbook 2011, China Statistics Press, Beijing, China, 2011.

29. National Bureau of Statistics, China Energy Statistical Yearbook 2011, China Statistics Press, Beijing, China, 2011.

30. IPCC, "IPCC Guidelines for National Greenhouse Gas Inventories," 2006.

31. K. Wang, X. Zhang, Y.-M. Wei, and S. Yu, "Regional allocation of CO_2 emissions allowance over provinces in China by 2020," Energy Policy, vol. 54, pp. 214–229, 2013.

32. F. Xi, Y. Geng, X. Chen et al., "Contributing to local policy making on GHG emission reduction through inventorying and attribution: a case study of Shenyang, China," Energy Policy, vol. 39, no. 10, pp. 5999–

6010, 2011.C. Kennedy, S. Demoullin, and E. Mohareb, "Cities reducing their greenhouse gas emissions," Energy Policy, vol. 49, pp. 774–777, 2012.

33. Shenzhen Municipal People›s Government, Shenzhen Low-Carbon Development Mid-to-Long Term Plan, Shenzhen Municipal People›s Government, Shenzhen, China, 2012.

34. S. E. Lee and G. J. Levermore, "Simulating urban heat island effects with climate change on a Manchester house," Building Services Engineering Research & Technology, vol. 34, no. 2, pp. 203–221, 2012.

35. J.-M. Feng, Y.-L. Wang, Z.-G. Ma, and Y.-H. Liu, "Simulating the regional impacts of urbanization and anthropogenic heat release on climate across China," Journal of Climate, vol. 25, no. 20, pp. 7197–7203, 2012.

36. V. Czako, "Drowning the suburb: settlement planning and climate change adaptation in a Hungarian metropolitan area," Urban Research & Practice, vol. 6, no. 1, pp. 95–109, 2013.

37. J. S. Risbey, "Dangerous climate change and water resources in Australia," Regional Environmental Change, vol. 11, no. 1, supplement, pp. 197–203, 2011.

Chapter 7

CULTIVATING A VALUE FOR NON-HUMAN INTERESTS THROUGH THE CONVERGENCE OF ANIMAL WELFARE, ANIMAL RIGHTS, AND DEEP ECOLOGY IN ENVIRONMENTAL EDUCATION

Helen Kopnina[1], and Brett Cherniak[2]

[1]Institute Cultural Anthropology and Development Sociology, Social and Behavioural Sciences, Leiden University, Wassenaarseweg 52, Leiden 2300 RB, The Netherlands
[2]47 Jones Street, Hamilton, Ontario, ON L8R 1X9, Canada

ABSTRACT

While the original objective of environmental education (EE) and education for sustainable development (ESD) acquired an awareness of the natural world and its current plight, animal welfare (AW), animal rights (AR), and deep ecology (DE) have often been absent within EE and ESD. AW and AR focus their attention on individual animals, while the DE perspective recognizes the intrinsic value of the environment. In this article, we shall discuss how the integration of these three approaches within EE/ESD can and should be improved, with particular reference to the ethical underpinnings of educational scholarship and practice. This article will argue that these three positions are well placed to enhance the democratic practices of EE/ESD through the adoption of an inclusive pluralism that embraces representation of non-human species and recognizes their interests.

INTRODUCTION

The objectives of environmental education (EE) are defined as helping students acquire: an understanding of basic ecology, an awareness of the natural world and its current plight, a sensitivity to the need for protecting nature, and the skills to help address environmental challenges [1]. Education for environmental sustainability engages both the macro units of study, such as "nature" or "ecosystems," and the micro, such as species and individual animals, plants, fungi, or bacteria. The scope of this article does not allow for a more detailed discussion of what qualifies as "animals" present in animal rights literature. For the purpose of this article, we shall only refer here to animals as sentient beings, thus a category that can also sometimes apply to plants, fungi, or bacteria. Simultaneously growing alongside EE over the decades, education addressing animal welfare and animal rights has emerged as part of a larger environmental concern [2].

The ethical underpinnings of animal rights (AR), animal welfare (AW) and deep ecology (DE) are based on a number of ethical positions relating to EE's original objectives. These ethical positions range from concern for animal welfare and animal rights on one hand to environmental ethics on the other. According to Tom Regan [3], animal rights are defined as a commitment to the abolition of the use of animals in science; dissolution of commercial animal agriculture and elimination of commercial and sport hunting and trapping. Environmental ethics, in turn, range from anthropocentrism (human-centeredness) to ecocentrism or biocentrism (placing ecosystems or biosphere at the heart of moral consideration) as well as the in-between gradations, including strong and weak anthropocentrism. Norton [4] distinguished between "felt" and "considered" preferences. A felt preference is one that may be temporarily satisfied by some specific experience. A considered preference is arrived at after "careful deliberations." According to Norton, an ethic is strongly anthropocentric if it focuses on felt preferences without reference to socio-cultural context (thus, non-consequentialism) and weak anthropocentrism if it focuses on felt and considered preferences, dependent on one's worldview (thus, consequentialism. Naess [5] drew a distinction between deep and shallow ecology, with the former associated with recognition of intrinsic value of environment beyond its unity, and the latter including human interests. While anthropocentrism assigns moral value to humans only, it can also include consideration of the environment when human welfare is affected, or of animals used for food, research, entertainment or companionship. Generally, non-anthropocentric ethical perspectives are concerned with the preservation of biodiversity and conservation—not so much for its utility to us humans, but for the benefit of entire habitats or species themselves. These general ethical orientations consider the moral position, welfare, or legal rights of either environment or individual animals, or both.

AW refers to the desire to prevent unnecessary animal suffering, and while not categorically opposed to the use of animals, wanting to ensure a good quality of life. AW includes not only the state of the animal's body, but also its feelings or sentience [6]. Animal welfare education (AWE) promotes knowledge, understanding, skills, attitudes and values related to human involvement in the lives of animals [7]. A number of educational programs and specialized courses have been established that address animal welfare from ethical, political, and practical perspectives. The Animal Welfare Institute (AWI), founded in 1951, and the International Fund for Animal Welfare (IFAW), founded in 1969, are both involved in education for animal rights and welfare [8,9]. However, their educational programs are not connected to EE nor education for sustainable development (ESD).

AR refers to the philosophical belief that animals should have the right to live their lives free of human intervention, opposing the use of some animals by humans [3]. In education, AR has not been taught independently from AWE, nor has "humane education", defined by the Institute for Humane Education as a form of education that "instills the desire and capacity to live with compassion, integrity, and wisdom, but also provides the knowledge and tools to put our values into action in meaningful, far-reaching ways."

Education for DE has been taught as part of outdoor (experience) education when the students are involved in outdoor activities [10] or philosophy courses [11]. AR, AW and DE have routinely been taught in existing courses such as biology, ethics, or sustainability, with independent focus being scarce (e.g., [12]). Within ESD, environmental sustainability concerns are normally displayed along with social and economic objectives, and AR, AW and DE are presented as an *ad hoc* practice, primarily associated with the conservationist and the naturalist subjects (e.g., [13,14,15,16]).

In this article, we shall address the ethical underpinning of the EE/ESD scholarship and reflect upon the position of AR, AW and DE within the established practice. We shall focus on the questions: How is care for ecosystems, biosphere, species or individual animals positioned within EE/ESD? What ethical positions underlie considerations of environment and animals in EE/ESD?

FROM ETHICS TO SUSTAINABILITY

The complexity of the natural world has necessitated the exploration and development of several nuanced and distinct non-anthropocentric ethics. Kronlid and Öhman [17] (p. 23) have distinguished between five different typologies of non-anthropocentric ethics in the context of EE research: sentientism (non-human animals with the ability to experience pain and suffering have intrinsic value); animal rights (non-human animals with a sense of self have intrinsic value); social animal ethics (non-human animals with the ability to engage in relationships have intrinsic value); biocentrism (all organisms have intrinsic value by virtue of having a good of their own related to their flourishing), and ecocentrism (ecosystems and species have intrinsic value because it is possible to relate to them as separate entities and because they have a capacity to sustain life and well-being in terms of their integrity, stability and beauty). These distinctions allow us to consider a wider range of perspectives, values and situations when discussing the position of non-humans within education practice while bringing to light the many concerns of the natural world.

Many theorists have attempted to demonstrate the precise ways in which deep ecology and animal rights/ethics/liberation are entangled and interdependent, as well as how they differ, for a review of some perspectives see [18]. AR, AW and DE are divided on a number of points but are united in their support of care for the environment and individual animals. The greatest commonality between DE and much of animal ethics is that they remove humans from the moral pedestal and seek recognition of the plight of non-humans. As Jamieson [19] has pointed out, the urgency of environmental predicament and the plight of non-humans does not allow for more internal battles, thus calling different "camps" to combine their knowledge, not to "overwhelm" the status quo with a flurry of "niche" groups.

This plight of non-humans is characterized primarily by the destruction of habitats upon which wildlife is dependent and the abuse of animals used for human consumption through the industrial food production system, e.g., concentrated animal feeding operations (CAFOs). Ecocentric scholars have argued that unless the intrinsic value of non-humans is recognized, there will be no institutional guarantees that the interests of other species will be protected and not continually neglected [20,21,22,23]. The market economy has been represented as the solution to issues of sustainability, embedding economic reasoning deeply into environmental policy, planning, and practice and ignoring the plight of the planet [24]. Such reasoning led to the subordination of non-humans and their habitats to social and economic objectives [1,13,17,25,26].

This trend of presenting nature as capital and commodity has made its way into educational practice. The literature is replete with references to natural *resources*, natural *capital*, and ecosystem *services*, conceptualizing nature through an anthropocentric, utilitarian lens while the recognition of the *intrinsic* value of biodiversity seldom appears in the same space. [1,13,17,27]. Environmental "management" of "natural resources" became interlinked with the market mechanisms of "species banking" and "carbon trading." In line with the Brundtland definition of "development that meets the needs of the present without compromising the ability of future generations to meet their own needs," ESD has shifted its focus towards environmental justice, primarily concerned with the distribution of environmental benefits and burdens among human beings [28]. ESD chiefly promotes sustainability, placing its focus on a "sense of justice, responsibility, exploration and dialogue" [29]. In this context, education concerned with animals is not commonly discussed as an integral part of education for sustainability. We shall argue that it is crucial to see AR, AW, and DE connected to the overall goal of enhanced empathy, which is stressed in social equity, as in Norton's convergence theory [4]. Convergence can be illustrated by the consumers' insistence on transparency in our food system, which is linked to the treatment of farm laborers as well as livestock; or address pollution that affects both human and natural systems. For example, as the World Society for the Protection of Animals (WSPA) has outlined, AWE entered ESD through the avenues of: environmental and agricultural sustainability (responsible animal management); human health (good animal care reduces the risk and spread of diseases that can be transmitted to humans and of food poisoning); poverty and hunger reduction (looking after animals properly improves their productivity and helps farmers to provide a secure food supply and income for themselves, helping to alleviate poverty); disaster preparedness and risk reduction (animals are important for people's lives and livelihoods and must be given due consideration in plans for disaster preparedness and response). It is notable that a view of "useful animals" is interlinked with conceptions of stewardship, management, and other optimistic "innovations" [30].

However, we also warn that when developing a framework for connecting AR, AW, and DE to sustainability education—as, for example, through ESD that emphasizes normative competencies, which include values and issues of morality—we need to be careful not to over-emphasize compromise and token accomplishments when harder choices about economic development *versus* preservation of biodiversity need to be made. "Improving animal productivity" has little relation to ethical concerns about non-humans, giving way to the broad aims embodied by ESD, such as social equality and economic equity (e.g., [31,32]). While there has long been a supposed connection between the conscientious consumption of green products and conservation, empirical evidence from wealthier countries denies that consumer responsibility alone can overcome unsustainability. Nor have multinational corporations shown much transformative power in their ability to alter global unsustainability practices or address the mere scale of animal use in industries ranging from food to pharmaceuticals [2,3,5]. Societal concerns do not naturally and automatically seep into educational practice, and neither can we assume that corporate or political leaders can be "taught" to care about non-humans. What can be achieved, however, is the critical awareness of how normative concepts of "sustainable development" or "animal productivity" are used. Yet, this critical reflection and skepticism about the indoctrinating tendencies of normative concepts should not turn into a fear of all types of instrumentality, as we shall explain in more detail below. Some EE/ESD scholars have expressed their angst of indoctrination leading to behavioral change that "environmentalists" could implant into their pupils' brains, even if this change means a more inclusive recognition of non-humans [33,34,35]. Illustrating the fear of environmentalism, they have warned against "eco-totalitarianism" [33] (p. 225). Breiting [34] has expressed concern with educators and pupils being used as marionettes by the generalized camp of "environmentalists". A telling example is provided by Bob Jickling. In reflecting upon the controversy about shooting wolves in the Yukon area, Jickling felt that advocating the pro-wolf position as a schoolteacher in a local community would be "neither practically viable nor educationally justifiable" [35] (p. 92). Jickling justifies this position by the need to stay neutral in order to teach students democratic and open values and avoid indoctrination.

Other EE/ESD scholars have tried to bridge the gap between the generalized camps of ecocentric environmentalists and "emancipatory" educators by arguing that educational practitioners and theorists should not continually reproduce a simplified notion of anthropocentrism and non-anthropocentrism [36]. Kronlid and Öhman [17] (p. 34) note that when environmental ethics is used in EE research, the "cross-disciplinary work *should* take the complexity and pluralism of environmental ethical issues and the variety of sub-positions produced above into consideration" (italics added). This may be seen as another example of the "latent authoritarian tendency" of environmentalists [37] (p. 126): the presumption that open discussions are guaranteed to solve all of our problems and produce the desired results. Rather than dwelling on the anthropocentrism/ecocentrism dichotomy, they suggest that educators should embrace different ethical perspectives and foster democratic learning.

The onus for drawing attention to AW and AR appears to have been left by educationists to outside pressure groups and the media [2], with DE being relegated to the marginal experiential education [10]. With a few exceptions, a couple of educational programs focus on intrinsic values or the rights of animals. This raises a number of ethical questions in both EE/ESD and within the general enterprise of sustainable development: Should future generations of humans have the right to experience wilderness and biological diversity? Does sustainable development concern other species? Do animals and plants have the right to a secure future? [36] (p. 67). Why is teaching for democracy, gender and racial equality and economic equity an expressed aim in most EE/ESD while the ongoing extinction or treatment of animals in the industrial food production system is a marginal concern?

AR, AW and DE have been present *ad hoc* for a few decades within or outside of EE or ESD, with issues concerning speciesism (discrimination against other species) seen as a footnote in mainstream environmental education journals (for review see [28]).

ANIMAL RIGHTS, ANIMAL WELFARE, AND DEEP ECOLOGY IN ENVIRONMENTAL ETHICS

Critics have pointed out that much of the anthropocentric ethics prevalent in EE/ESD is primarily concerned with questions of social and economic equity [28]. "Social equity," for example, usually means "universal participation in global markets that will require continued ecological sacrifices" (e.g., [25]). In fact, the moral imperatives of social and economic equality conveniently serve the global market through the global perpetuation of consumer culture [26]. Considerations of non-humans are marginal in the pluralistic perspective dominated by social and economic agendas [27,28].

Will we even notice when species go extinct, and if we do, what will we notice? Indeed, there are plenty of empirical examples which demonstrate that the extinction of species have not affected food supply chains or human welfare. As witnessed during recent extinctions, few direct negative side effects are experienced by people, aside from some marginal benefits for the pharmaceutical industry or zoos. Thompson [38] and Kareiva *et al.* [39] have found that we do not need to save every species as humanity is not dependent on them, but should instead embrace human-managed nature and no longer waste time teaching students to care about what is left of the wilderness. This transition of academic discourse toward human-managed nature is part of a more general trend that embraces the domestication of the planet. Yet critics have argued that this instrumental attitude to nature is precisely what has brought us to our current environmental problems [27].

While many different schools of environmental ethicists agree that humans and environment are intimately interrelated, interconnected, and the relationship between the human and non-human worlds is that of mutual reciprocity, more skeptical critics emphasize that this reconciliatory view tends to ignore how economic development destroys entire habitats and distracts public attention from the plight of farm animals or those used for medical experimentation [25]. According to this critique, it is not a question of whether humans and nature are connected or dependent on nature (in a similar way, we could inquire whether the slave owners were really dependent on slaves), but whether these elements of nature—living creatures—have any significant moral status. Is there not a moral need to recognize that we are not the only species on Earth that needs to be considered? To return to the question posed in the introduction, what ethical positions underlie considerations of the environment and animals in EE/ESD?

THE QUESTION OF MORALITY

While there is a shift in public awareness towards recognition of animal welfare concerns, there is no consistent discussion about the *scale* of instrumental use of other species. The scale of this exploitation, including CAFOs and intensive farming, which relegate billions of animals to short and miserable existences, has increased exponentially alongside human population growth and consumption [25]. While the fate of a single abused dog may capture public attention through the media, there is no consistent discussion about the millions of species "harvested" for consumption, or used for medical experiments.

The philosophical and political difficulty of establishing a non-anthropocentric ethic has led pragmatic theorists to support an "enlightened anthropocentrism" which, whilst recognizing the human-centered reality, claims to be consistent with a high degree of protection for non-human nature (e.g., [4,40,41]). Such a case is based primarily on an empirical claim that the protection of the natural world—including animals—is in the interest of humans. However, the pragmatist thinkers reflect that relying on the intrinsic value of non-human animals and the environment is not sustainable and argue that moral anthropocentrism is unavoidable [41,42]. Enlightened anthropocentrism is largely consistent with "the convergence thesis," stating that preservationist and conservationist policies will tend to converge in the long run [43].

For example, biological and organic meat can be good both for farm animals and for human health. Another familiar example is addressing water, soil or air pollution that is likely to harm humans with the positive side-effects it brings for habitats or species. These lead to the type of convergence thinking that assumes a healthy environment caters to human needs and, conversely, that social or economic needs will serve nature. This reliance on perceived interdependence is often seen in EE/ESD. Famously, the Brundtland report speaks of convergence between the three pillars of economic development, social equity, and environmental protection. A similar effort is made to "balance" opposing perspectives, presenting conflicts not in moral (good *vs.* bad) or rational (right *vs.* wrong) terms but as a broader exercise in social learning [44] or "learning from sustainable development" [45].

However, others have noted that occasionally the interests of humans and animals (and humans and environment more broadly) coincide, anthropocentrism can make a positive contribution to the evaluation and justification of environmental policy in situations dealing with artificial (human-created or controlled) and not natural systems [46]. Examples of policies addressing artificial systems might include urban air pollution, regulation of greenhouse gases, issues of environmental justice, and the environmental impacts of agriculture. These are largely consistent with conventional sustainability policy, where the environment and individual species or animals are kept in orbit with economics at the center; our destructive systems become more sustainable, leading us to believe that we are helping the environment as a whole. The environment in this policy will be considered if and only if a degrading environment might undermine ongoing development [47]. However, in cases of largely natural systems, wilderness preservation and the protection of endangered species, a moral monism [48] embracing ecocentric, biocentric or animal-centric positions offers a wider protection for environment than an anthropocentric-dominant pluralism. According to Katz "there is a clear difference between anthropocentric and nonanthropocentric justifications for environmental policy, and in these situations I remain committed (mercilessly) to nonanthropocentrism" [46] (p. 390).

Naess argued that sustainability hinges on a consistent education curriculum which is tied back to a well-informed understanding of the state of the planet [49]. The non-anthropocentric environmental ethics calls for a new ethic, not beholden to any previously dominant stream of thought and not dependent on the shifting cultural morality of today [19]. An imperative to protect and preserve (or, to use a more popular term, to accept stewardship of) non-humans could stem from human reason, or love, or sense of duty and responsibility [47].

Yet, this ethic is now in short supply if we notice how mercilessly the processes endangering habitats, species and individual animals have expanded their hold within public imagination, and that of EE/ESD scholars. The critics have noted that the brainwashed, anthropocentric ideology that has taken hold of the supposedly free "choices" being made in neoliberal education systems is not likely to offer a way forward as the choices given still find themselves under the influence of this anthropocentric ideology [24,50,51].

REFLECTION

In answering the question, "How is care for ecosystems, biosphere, species or individual animals positioned within EE/ESD?" we note that this possibility is contingent upon a socio-political and cultural context permitting an ecocentric ethic to be recognized as a necessary contribution to environmental integrity. An EE/ESD that is already respective and inclusive of AW, AR and DE could help to create this context. As with other socio-political changes, education plays an essential role in change to society as a whole and the political institutions within. Still, we need to be aware that if inroads are only made in education, this progress might be lost once students leave the classroom if the greater context is not in place, thus a wider societal transformation that offers an ethical support base for AW, AR and DE would be close to a requirement. An education is pointless if students enter a society that does not allow for the inclusive support of the values and perspectives they have learned.

Encouragingly, Andrew Dobson has suggested that non-human animals could be democratically represented through proxy representatives elected explicitly to promote their interests, occurring through real elections in which candidates are elected in the present system [22]. In his later work, Dobson argued that since animals cannot speak themselves, the ability to recognize their "voice" through their representatives, is of crucial importance if they are to be considered as democratic (with "demos" in this case extending to non-humans) participants [23]. In this case we need to note that the voice of animals (as well as fungi, algae, *etc.*) can be extended to habitats and ecosystems by combining common threads from AR, AW and DE. Generally, these common threads are based on the philosophy of inclusion of ecosystems, species or non-human individuals within the sphere of moral consideration, based on biophilia (love of nature), and, particularly in the case of DE, realization of intimate interconnectedness between humans and natural elements.

While DE's philosophy of inclusion might be too weak in the face of dominant anthropocentric neo-liberal ideology, AR might be too "strong." Presently, AR is often perceived as subversive or threatening. For example, Animal Liberation Front (ALF) and Earth Liberation Front (ELF) activists are seen as dangerous and radical [52,53,54]. While such organizations exhibit a concern that is desirable if non-human interests are to be upheld, they do not address the social, economic, or cultural forces that shape the general public's attitudes. However, it is still precisely this type of "radical" attitude that is needed to reverse the impetus of the presently taken-for-granted anthropocentric ethics that governs much of EE/ESD educational practice. This is not to imply that the acts of property destruction that these activists have committed are recommended for educators to teach to students, but rather understanding the type and development of commitment, rage, as well as selfless love of non-human beings, be they clear-cut forests or animals used for laboratory experiments that deserve educators' attention. Richard Kahn discussed the need of radical reorientation of education toward ecological concerns and other species, examining how the essence of ALF and ELF's ideas (although certainly not the methods) can be used to inspire educational practice beyond the current status quo [53]. Integrating the perspectives from groups such as ELF in the curriculum was discussed by Kopnina [55,56]. It was suggested that we can draw inspiration for educational practice from eco-pedagogy, which places ecological integrity as well as animal welfare at the center of sustainability challenges. An example of this is given by Kopnina in her discussion of exposing business students to radical environmental activism through the in-class viewing of the documentary film, *If a Tree Falls* (2011), directed by Marshall Curry and Sam Cullman about the rise and fall of ELF. As one of the reviewers of the film wrote:

What is interesting is that even while we don't agree with what the ELF is doing, the film gives us images that allow us to understand their point of view. We see footage of trees that have stood for thousands of years, blindly cut down. We see horse mills, with hundreds of dead horses hung from the ceiling. We see the heartbreaking sight of a group of legendary trees sawed down to make a parking lot... We see the protesters themselves, camped out in the trees that are to be cut down, beaten unmercifully by the local police... The irony is that the members of the group who are clearly guilty of vandalism haven't done any physical harm to other human beings but are being beaten down by law enforcement as if they were murderers [57]

After viewing the film, the business students exhibited a shift in their ethical thinking and an increase in awareness of ecocentric concerns. The film made students question the activists' reasons behind their level of commitment, contemplating what makes some people join ALF and ELF and others refrain. In written assignments following the viewing of the film, students were able to separate the beliefs of ALF/ELF from the stigma attached to their actions. The students realized that the "radical" or "terrorist" label of the activism featured in the film could indeed be disputed—to quote a law enforcement officer in the film, "One man's terrorist is another man's freedom fighter."

More generally, returning to educational practice and questions which researchers should explore further: Do we need their "strong" views throughout the entire process, or just to begin "the shift"? What creates a positive feeling about non-humans, and how can we educators help to spread that? Would aligning EE/ESD with some of the aims of radical organizations, or only their attitudes, provide an easy target for opposition?

This shift in awareness is also crucial in relation to AR and AW. Regarding animal ethics and welfare in education, Kelly reflected that "those who know it is not necessary to kill animals for food or fight them as entertainment have an ethical responsibility to teach other humans how to eat without killing, how to be entertained without exploiting" [58]. According to Kelly, this type of education triggering a shift in awareness and ethical consideration of non-humans would "allow researchers to devote research to making animals' lives better without worrying about the accusation that they should be devoting their time to less trivial matters; and would function to make education a far less anthropocentric endeavor" [58] (p. 45). This research could form a base, noted Kelly, to teaching about legal protection of animals that (1) emphasize respect for animals, even when others understand this as moral proselytizing; (2) teach that it is wrong to continue practices and habits that are exploitive of animals when alternative practices are possible; (3) design life-skills curricula that help those who cannot envision the feasibility of these alternative practices to do so (such as vegetarian cooking); (4) de-anthropocentrize the mission of educational institutions; and (5) discount any unnecessary piece of knowledge gained at the expense of animals' welfare as unworthy of academic attention [58] (p. 45). These steps will require continuous representation of non-human interests by eco-advocate educators.

Thus, the role of representation of non-human interests through eco-advocates [59,60] is essential if educational researchers and practitioners are to move beyond the status quo of conventional plurality. In discussing environmental justice (equal distribution of environmental risks and benefits, including to the non-human species) and democracy, Dobson emphasized that "if harm is being done, then more justice rather than more talking is the first requirement" [22] (p. 26).

Because other species cannot engage in "pluralistic" discussion due to their inability to speak our language, the participation of educators who represent non-human "voices" is essential. Following this, the insistence that the distinction between anthropocentric and non-anthropocentric ethics is not relevant for EE/ESD (e.g., [17,36]) is counterproductive to the effort to expand the pedestal of moral significance to the non-humans. Questioning anthropocentrism is far more than an academic exercise of debating the dominant cultural motif of placing humans at the center of material and ethical concerns. It is a fertile way of shifting the focus of attention away from the problem-symptoms of our time to the investigation of root causes. And certainly the dominant beliefs, values, and attitudes guiding human action constitute a significant driver of the pressing problems of our day [61].

The misplaced fear of eco-totalitarianism in education [33] fails to account for the one-species' totalitarianism over non-human beings, denying eco-advocates the right of providing the "voice" to the millions in need of urgent protection. The dominant form of pluralism is often reflective of the neoliberal model in which individual choices are intertwined with free market thinking at the expense of ecological concerns [50,51]. Anthropocentric pluralism places humanity outside and above the natural world, and its constructed distinction and artificial perch offer the illusion of superiority and a transient experience of "wealth" at the expense of others, while what is sacrificed is the very source from which true power flows: abundance, creativity, unexpectedness, reciprocity, and mutual flourishing [61]. One-species-only pluralism leads educators to think that they give students choices to choose plural perspectives, while the choices are between different shades of anthropocentrism.

Inclusive pluralism embraces non-human representation, while human eco-advocates "speak for nature" [62] and billions of Earth's citizens. As a starting point, we could discuss what these non-human voices are exactly saying, and whether some should count more than others, and why some educators should claim to speak for them. Without presuming to know the answer to either of these questions, we support one simple claim: that sentient beings, from orangutans to plants, want to survive, and it is our shared responsibility as members of one single species that both endangers and can act on behalf of their survival, to consider them in our 'pluralist' ethics and education. An inclusive pluralism recognizes plurality of choices of *all* Earth's citizens, not just those of one species. Andrew Dobson's work in relation to democratic inclusion of non-humans [22,23] as well as Arne Naess's work [5,49] on how different forms of respect and inclusion of non-humans might be grounded provides hopeful directions for consequent research on inclusive pluralism in educational contexts.

However, we realize that incorporating environmental ethics, animal welfare and animal rights into a seamless synthesis easily taught to students can be difficult. There are great conceptual difficulties involved in extending ethics beyond its current application. The strength of animal ethics lies not in incorporating it as part of environmental ethics, which itself seems difficult to subsume under current societal ethics. How does one extend our social ethic to non-sentient and abstract entities such as wilderness areas or even species? The extension of ethics to animals was accomplished in the 70's by various philosophers including Singer and Rollin when they demonstrated the absence of morally relevant differences between humans and animals, and the presence of morally relevant similarities. It is in fact a far more difficult task to extend moral status to entities not continuous in traits with humans. Yet, if we despair, no change is likely to come.

HOPEFUL DIRECTIONS

Examples of educational practice that integrate inclusive pluralism in which non-human agents are recognized as potential contributors to diversity perspectives include conservation education [63,64,65], outdoor education [66,67], education for deep ecology (e.g., [10,11,68]), AWE [2] and post-humanist education [69,70]. All these different directions offer a hope of restoration—not just that of nature and animal welfare, but also of ourselves as caring educators and students. As Richard Louv [71] (p. 226), the author of a popular book *Last Child in the Woods* has stated, "An environment-based education movement—at all levels of education—will help students realize that school isn't supposed to be a polite form of incarceration, but a portal to the wider world" and, as this article makes clear, that wider world includes the inner world. This wider world includes the possibility that lecturers could transmit to their students the ability to transcend the top-down sustainable development that is usually based on control rather than compassion. Just as many conservationists are more interested in increasing their capacity for empathy than for control, they are also frequently interested in communicating to larger audiences how and why their knowledge of nature leads them to love nature [72]. In the words of Vucetich and Nelson [72] (p. 13), knowing conservation›s role in society requires knowing how and why populations and ecosystems are valuable. In particular, we need to know how they are valuable beyond their utility to humans—and this is the task that EE and ESD are well suited to undertaking. However, EE/ESD need to be reoriented toward a clearer distinction between what is socially and economically defined as sustainability and what is ecologically defined sustainability—in other words, the difference between sustainable development as a slogan that implies sustaining economic growth, and environmental sustainability that underlies long-term survival of social, economic and natural systems [73].

A hopeful direction for non-anthropocentric education includes initiatives by a number of charities and foundations. The Foundation for Deep Ecology (FDE) supports education and advocacy on behalf of wild Nature, educating citizens that stopping the global extinction crisis and achieving true ecological sustainability will require rethinking our values as a society. FDE argues that "present assumptions about economics, development, and the place of human beings in the natural order must be reevaluated." FDE's educational programs support restoration of "traditional knowledge, values, and ethics of behavior that celebrate the intrinsic value and sacredness of the natural world and give the preservation of Nature prime importance," thereby introducing students to broader topics of sustainability with nature and animals as focal points [74]. In a similar effort, the Rewilding Foundation [75] participates in ecological education and participatory citizen science projects.

Another hopeful direction is integrating AR, AW, and DE within national platforms of EE and ESD, as, for example, the inclusion of Animals (with subsections specified e.g., "Animal welfare", and "Endangered species") by the National Association for Environmental Education UK [76]. At present, these topic sections are empty, implying that it is only a gesture of good will. Also in England, charities like the Royal Society for the Prevention of Cruelty to Animals (RSPCA) offer free training and support for professionals who work with children and young people, stating that their educational mission is to help students to develop informed, responsible and active citizens [77].

Tellingly, RSPCA combines the elements of AR, AW and DE by offering both domestic animal-oriented as well as wilderness-oriented curricula through their cooperation with animal shelters, an outdoor education center and a wildlife hospital. People for the Ethical Treatment of Animals (PETA) is the largest animal welfare organization in the world and also allocates education a special role, linking the values encompassing respect for animals, nature, and other human beings, as well as active citizenship:

Teaching kids to have compassion and empathy for their furry, feathered, and finned friends is vital for preventing cruelty to animals as well as in raising them to respect and treat those who are different from them with kindness [78]

These are some among many promising paths that integrate true pluralism by which we mean inclusion of all species into moral consideration and responsible citizenship within and beyond EE and ESD.

CONCLUSIONS

In this article, it was argued that education for sustainability should consider integrating animal rights, animal welfare, and deep ecology (DE) into the core of environment education (EE) and education for sustainable development (ESD) and that these three directions should guide education for sustainability. The instrumental attitude to nature appears to be insufficient for protection of the most vulnerable elements of the environment, or for addressing animal welfare or animal rights, and raising ethical objectives to the anthropocentric view of nature in education. Conservation education, education for deep ecology and education for animal rights can be seen as education that embraces a *truly plural* position that recognizes diversity of *all* species. This position would entail a critical education exposing the deficiencies of the mainstream anthropocentric ethics. This might necessitate a continuous affirmative action program as, unlike disenfranchised humans, non-humans will never be able to speak, even when threatened with extinction.

Mike Appleby, an animal welfare scientist was once asked: "What should we do about animal welfare?" He replied: "The answer, ladies and gentlemen, is 'More'." [79]. The same could be said about the question of what should be done about ethical considerations of environment and animals in EE and ESD [60]. What precise shape such recognition will take in both societal and educational discourse still needs to be discussed (e.g., should we oppose all research on animals, or only that which is particularly cruel and not for any meaningful purpose, *i.e.*, cosmetics?). The need for this urgent discussion should be a starting point. While we as researchers and academics would certainly benefit from extended discussion of how the non-human voices can be best represented and by whom, the move toward recognition that the very survival of these non-humans is at stake deserves to be recognized in education for sustainability. This type of change will require affirmative action programs on behalf of what is left of wild Nature and of a wonderful human capacity for empathy and compassion.

AUTHOR CONTRIBUTIONS

Helen Kopnina conceived and set out the initial purpose of the work. Helen Kopnina and Brett Cherniak analyzed the relevant material. Helen Kopnina provided extensive experience in environmental education and animal rights education research. Brett Cherniak contributed critical discourse and experience in the democratic element of environmental education. Helen Kopnina and Brett Cherniak thoroughly edited the article throughout its development and review process. Helen Kopnina and Brett Cherniak agreed and reviewed upon all changes before any round of submissions. Helen Kopnina and Brett Cherniak wrote the paper.

REFERENCES

1. UNESCO. *The Belgrade Charter*; UNESCO: Paris, France, 1976.
2. Marsden, W.E. Animal welfare issues in geography and environmental education: Some British historical perspectives. *Int. Res. Geogr. Environ. Educ.* 2001, *10*, 394–410.
3. Regan, T. A Case for Animal Rights. In *Advances in Animal Welfare Science*; Fox, M.W., Mickley, L.D., Eds.; The Humane Society of the United States: Washington, DC, USA, 1986; pp. 179–189.
4. Norton, B.G. Environmental ethics and weak anthropocentrism. *Environ. Eth.* 1984, *6*, 131–148.
5. Naess, A. The shallow and the deep: Long-range ecology movement. *Inquiry* 1973, *16*, 95–99.
6. Hewson, C. What is animal welfare? Common definitions and their practical consequences. *Can. Vet. J.* 2003, *44*, 496–499.
7. World Society for the Protection of Animals. Education for Sustainable development. Available online: http://www.animalmosaic.org/Images/Education%20for%20Sustainable%20Development_tcm46-29460.pdf (accessed on 21 November 2015).
8. Animal Welfare Institute. Humane education. Available online: https://awionline.org/content/humane-education(accessed on 21 November 2015).
9. International Fund for Animal Welfare. Available online: http://www.

ifaw.org/united-states/our-work/education(accessed on 21 November 2015).

10. LaChapelle, D. Educating for deep ecology. *J. Exp. Educ.* 1991, *14*, 18–22.

11. Drengson, A.R. Introduction: Environmental crisis, education, and deep ecology. *Trumpeter* 1991, *8*, 97–98.

12. Root-Bernstein, M.; Root-Bernstein, M.; Root-Bernstein, R. Tools for thinking applied to nature provide an inclusive pedagogical framework for environmental education. *Fauna Flora Int. Oryx* 2014, *3*, 1–9.

13. Sauvé, L. Currents in environmental education: Mapping a complex and evolving pedagogical field. *Can. J. Environ. Educ.* 2005, *10*, 11–37.

14. Glasser, H. Learning Our Way to a Sustainable and Desirable World: Ideas Inspired by Arne Naess and Deep Ecology. In *Higher Education and the Challenge of Sustainability: Problematics, Promise, and Practice*; Corcoran, P.B., Wals, A.E.J., Eds.; Springer Netherlands: New York, NY, USA, 2004; pp. 131–148.

15. Gorski, P.C. Critical ties: The animal rights awakening of a social justice educator. 2009. Available online: http://www.edchange.org/publications/animal-rights-social-justice.pdf (accessed on 21 November 2015).

16. Hickman, J.A. Personal care products council: An animal cruelty analysis. 2010. Available online: http://www.edchange.org/multicultural/humane/pcpc.pdf (accessed on 21 November 2015).

17. Kronlid, D.O.; Öhman, J. An environmental ethical conceptual framework for research on sustainability and environmental education. *Environ. Educ. Res.* 2013, *19*, 21–44.

18. Garner, R. Environmental Politics, Animal Rights and Ecological Justice. In *Sustainability: Key Issues*; Kopnina, H., Shoreman-Ouimet, E., Eds.; Routledge Earthscan: New York, NY, USA, 2015.

19. Jamieson, D. Animal liberation is an environmental ethic. *Environ. Values* 1998, *7*, 41–57.

20. Eckersley, R. Liberal democracy and the rights of nature: The struggle for inclusion. *Environ. Polit.* 1995, *4*, 169–198.

21. Eckersley, R. *The Green State: Rethinking Democracy and Sovereignty*; MIT Press: London, UK, 2004.

22. Dobson, A. *Citizenship and the Environment*; Oxford University Press: Oxford, UK, 2003.

23. Dobson, A. *Listening for Democracy*; Oxford University Press: Oxford, UK, 2014.

24. Washington, H.G. *Human Dependence on Nature*; Routledge: New York, NY, USA, 2013.
25. Crist, E. Ecocide and the extinction of animal minds. In *Ignoring Nature no More: The Case for Compassionate Conservation*; Bekoff, M., Ed.; Chicago University Press: Chicago, IL, USA, 2013; pp. 45–53.
26. Crist, E. Abundant Earth and Population. In *Life on the Brink: Environmentalists Confront Overpopulation*; Cafaro, P., Crist, E., Eds.; University of Georgia Press: Athens, GA, USA, 2012; pp. 141–153.
27. Bonnett, M. Sustainable development, environmental education, and the significance of being in place. *Curr. J.* 2013, *24*, 250–271.
28. Kopnina, H. Education for sustainable development (ESD): The turn away from "environment" in environmental education? *Environ. Educ. Res.* 2012, *18*, 699–717.
29. Nevin, E. Education and sustainable development. *Policy Pract. A Dev. Educ. Rev.* 2008, *6*, 49–62.
30. Jickling, B.; Wals, A.E.J. Globalization and environmental education: Looking beyond sustainable development. *J. Curric. Stud.* 2008, *40*, 1–21.
31. UNESCO. United Nations Decade of Education for Sustainable Development (2005–2014). In *Framework for the International Implementation Scheme*; UNESCO: Paris, France, 2005; Volume 32C.
32. Wals, A.E.J. *Shaping the Education of Tomorrow: 2012 Full-Length Report on the UN Decade of Education for Sustainable Development*; UNESCO: Paris, France, 2012.
33. Wals, A.E.J.; Jickling, B. "Sustainability" in higher education: From doublethink and newspeak to critical thinking and meaningful learning. *Int. J. Sustain. High. Educ.* 2002, *3*, 221–232.
34. Breiting, S. Issues for environmental education and ESD research development: Looking ahead from WEEC 2007 in Durban. *Environ. Educ. Res.* 2009, *15*, 199–207.
35. Jickling, B. Wolves, Ethics and Education: Looking at Ethics through the Yukon Wolf Conservation and Management Plan. In *A Colloquium on Environment, Ethics and Education*; Jickling, B., Ed.; Yukon College: Whitehorse, YT, Canada, 1996; pp. 158–163.
36. Öhman, J.; Östman, L. Clarifying the ethical tendency in education for sustainable development practice: A wittgenstein-inspired approach. *Can. J. Environ. Educ.* 2008, *13*, 57–72.
37. Torgerson, D. *The Promise of Green Politics: Environmentalism and the*

Public Sphere; Duke University Press: Durham, NC, USA, 1999.

38. Thompson, K. *Do We Need Pandas? The Uncomfortable Truth about Biodiversity*; UIT Cambridge Ltd.: Cambridge, UK, 2010.

39. Kareiva, P.; Lalasz, R.; Marvier, M. Conservation in the anthropocene: Beyond solitude and fragility. *Breakthr. J.* 2011, 27–29. Available online: http://thebreakthrough.org/index.php/journal/past-issues/issue-2/conservation-in-the-anthropocene (accessed on 21 November 2015).

40. Barry, J. *Rethinking Green Politics: Nature, Virtue and Progress*; Sage: London, UK, 1999.

41. Hui, K. Moral anthropocentrism is unavoidable. *Am. J. Bioeth.* 2014, *14*, 25.

42. Light, A. Compatibilism in Political Ecology. In *Environmental Pragmatism*; Light, A., Katz, E., Eds.; Routledge: New York, NY, USA, 1996.

43. Norton, B.G. Conservation and preservation: A conceptual rehabilitation. *Environ. Eth.* 1986, *8*, 195–220.

44. Wals, A.E.J. Between knowing what is right and knowing that it is wrong to tell others what is right: On relativism, uncertainty and democracy in environmental and sustainability education. *Environ. Educ. Res.* 2010, *16*, 143–151.

45. Van Poeck, K.; Vandenabeele, J. Learning from sustainable development: Education in the light of public issues.*Environ. Educ. Res.* 2012, *18*, 541–552.

46. Katz, E. A pragmatic reconsideration of anthropocentrism. *Environ. Eth.* 1999, *21*, 377–390.

47. Rolston, H., III. Environmental Ethics for Tomorrow: Sustaining the Biosphere. In *Sustainability: Key Issues*; Kopnina, H., Shoreman-Ouimet, E., Eds.; Routledge Earthscan: New York, NY, USA, 2015.

48. Callicott, J.B. The pragmatic power and promise of theoretical environmental ethics. *Environ. Values* 2002, *11*, 3–25.

49. Naess, A. Beautiful action. Its function in the ecological crisis. *Environ. Values* 1993, *2*, 67–71.

50. Bansel, P. Subjects of choice and lifelong learning. *Int. J. Qual. Stud. Educ.* 2007, *20*, 283–300.

51. Davies, B.; Bansel, P. Neoliberalism and education. *Int. J. Qual. Stud. Educ.* 2014, *20*, 247–259.

52. Liddick, D.R. *Eco-terrorism: Radical Environmental and Animal*

Liberation Movements; Praeger Publishers: Santa Barbara, CA, USA, 2006.

53. Kahn, R. *Critical Pedagogy, Ecoliteracy and Planetary Crisis: The Ecopedagogy Movement*; Peter Lang: New York, NY, USA, 2010.
54. Kopnina, H. The Lorax complex: Deep ecology, ecocentrism and exclusion. *J. Integr. Environ. Sci.* 2012, *9*, 235–254.
55. Kopnina, H. If a tree falls: Business students' reflections on environmentalism. *Int. J. Environ. Sustain. Dev.* 2014, *8*, 311–329.
56. Kopnina, H. If a tree falls and everybody hears the sound: Teaching deep ecology to business students. *J. Educ. Sustain. Dev.* 2015, *9*, 1–16.
57. IMDB Film Reviews. If a Tree Falls: A Story of the Earth Liberation Front. Available online: http://www.imdb.com/title/tt1787725/reviews (accessed on 21 November 2015).
58. Kelly, S. Geschlecht, speciesism, and animal rights in Leopold von Sacher-Masoch. *J. Crit. Anim. Stud.* 2014, *12*, 28–50.
59. Lidskog, R.; Elander, I. Addressing climate change democratically. Multi-level governance, transnational networks and governmental structures. *Sustain. Dev.* 2009, *18*, 32–41.
60. Kopnina, H.; Gjerris, M. Are some animals more equal than others? Animal rights and deep ecology in environmental education. *Can. J. Environ. Educ.* 2015, *20*, 109–123.
61. Crist, E.; Kopnina, H. Unsettling anthropocentrism. *Dialect. Anthropol.* 2014, *38*, 387–396.
62. O'Neill, J. Who Speaks for Nature? In *How Nature Speaks: The Dynamics of the Human Ecological Condition*; Haila, Y., Dyke, C., Eds.; Duke University Press Books: Durham, NC, USA, 2006.
63. Norris, K.S.; Jacobson, S.K. Content analysis of tropical conservation education programs: Elements of success. *J. Environ. Educ.* 1998, *30*, 38–44.
64. Goodall, J. Caring for People and Valuing Forests in Africa. In *Protecting the Wild: Parks and Wilderness, The Foundation for Conservation*; Wuerthner, G., Crist, E., Butler, T., Eds.; The Island Press: Washington, DC, USA; London, UK, 2015; pp. 21–26.
65. Noss, R.F.; Dobson, A.P.; Baldwin, R.; Beier, P.; Davis, C.R.; DellaSala, D.A.; Francis, J.; Locke, H.; Nowak, K.; Lopez, R.; *et al.* Bolder Thinking for Conservation. In *Protecting the Wild: Parks and Wilderness, The Foundation for Conservation*; Wuerthner, G., Crist, E., Butler, T., Eds.; The Island Press: Washington, DC, USA; London, UK, 2015; pp. 17–21.

66. Sandell, K.; Öhman, J. Educational potentials of encounters with nature: Reflections from a Swedish outdoor perspective. *Environ. Educ. Res.* 2010, *16*, 113–132.

67. Galen D'Amato, L.; Krasny, M.E. Outdoor Adventure Education: Applying Transformative Learning Theory to Understanding Instrumental Learning and Personal Growth in Environmental Education. *J. Environ. Educ.* 2011, *42*, 237–254.

68. Glasser, H. Naess's deep ecology: Implications for the human prospect and challenges for the future. *Inq. Interdiscip. J. Philos.* 2011, *54*, 52–77.

69. Bonnett, M. Retrieving nature: Education for a post-humanist age. *J. Philos. Educ.* 2003, *37*, 551–730.

70. Bonnett, M. Sustainability, the Metaphysics of Mastery and Transcendent Nature. In *Sustainability: Key Issues*; Kopnina, H., Shoreman-Ouimet, E., Eds.; Routledge Earthscan: New York, NY, USA, 2015.

71. Louv, R. *Last Child in the Woods: Saving Our Children from Nature-Deficit Disorder*; Workman Publishing Company: New York, NY, USA, 2008.

72. Vucetich, J.A.; Nelson, M.P. The Infirm Ethical Foundations of Conservation. In *Ignoring Nature No More: The Case for Compassionate Conservation*; Bekoff, M., Ed.; University of Chicago Press: Chicago, IL, USA, 2013; pp. 9–25.

73. Washington, H. *Demystifying Sustainability: Towards Real Solutions*; Routledge: London, UK, 2015.

74. Foundation for Deep Ecology. Available online: http://deepecology.org/mission.htm (accessed on 23 November 2015).

75. Rewilding Foundation. Available online: www.rewildingfoundation.org (accessed on 23 November 2015).

76. National Association for Environmental Education (UK). Available online: http://naee.org.uk/ (accessed on 23 November 2015).

77. Royal Society for the Prevention of Cruelty to Animals. Available online: http://education.rspca.org.uk/education/home (accessed on 23 November 2015).

78. PETA Kids. Available online: http://www.petakids.com/parents/teaching-compassion/#ixzz3LXBoruUv (accessed on 21 November 2015).

79. Hewson, C. Focus on animal welfare. *Can. Vet. J.* 2003, *44*, 335–336.

Chapter 8

AN ECOLOGY FOR CITIES: A TRANSFORMATIONAL NEXUS OF DESIGN AND ECOLOGY TO ADVANCE CLIMATE CHANGE RESILIENCE AND URBAN SUSTAINABILITY

Daniel L. Childers[1], Mary L. Cadenasso[2], J. Morgan Grove[3], Victoria Marshall[4], Brian McGrath[4] and Steward T. A. Pickett[5]

[1]School of Sustainability, Arizona State University, Tempe, AZ 85287, USA
[2]Department of Plant Sciences, University of California, Davis, CA 95616, USA
[3]U.S. Forest Service, Baltimore, MD 21201, USA
[4]Parsons The New School for Design, New York, NY 10011, USA
[5]Cary Institute of Ecosystem Studies, Millbrook, NY 12545, USA

ABSTRACT

Cities around the world are facing an ever-increasing variety of challenges that seem to make more sustainable urban futures elusive. Many of these challenges are being driven by, and exacerbated by, increases in urban populations and climate change. Novel solutions are needed today if our cities are to have any hope of more sustainable and resilient futures. Because most of the environmental impacts of any project are manifest at the point of design, we posit that this is where a real difference in urban development can be made. To this end, we present a transformative model that merges urban design and ecology into an inclusive, creative, knowledge-to-action process. This design-ecology nexus—an ecology *for* cities—will redefine both the process and its products. In this paper we: (1) summarize the relationships among design, infrastructure, and urban development, emphasizing the importance of joining the three to achieve urban climate resilience and enhance sustainability; (2) discuss how urban ecology can move from an ecology *of* cities to an ecology *for* cities based on a knowledge-to-action agenda; (3) detail our model for a transformational urban design-ecology nexus, and; (4) demonstrate the efficacy of our model with several case studies.

INTRODUCTION

Sustainability is a broad ideal. It links social, environmental, and economic integrity through a process of democratic, inclusive goal setting [1,2]. Further, sustainability is a process, not an outcome or endpoint [3] and the transitions that cities undergo as they move towards more sustainable futures can take many forms [4]. In this paper we propose that an important way to enhance both urban sustainability and climate change resilience is for mutual learning between urban design practice and ecological science, with particular attention to adaptive processes and pluralistic governance. We define "design" as far more than a plan or drawing of an object or space before it is built or made. We conceive design to be the purpose, planning, or intention that exists, or is thought to exist, behind any action or object. It is a conscious effort to create order in all human activity and it includes our perceptions, actions, and review of the impacts of those actions (*sensu* [5]). Design can play a substantial role in environmental outcomes. For instance, Thackara [6] estimated that 80% of the environmental impact of any object, including the built environment, is determined at the design stage. Understanding and mitigating that impact demands a process of public deliberation that better integrates design and ecology in order to achieve sustainable urban development.

Throughout the paper, we focus our discussion and examples on challenges that cities face, and will face in the future, associated with climate change—with special emphasis on heat extremes, droughts, and floods. In the case of heat extremes, the urban heat island effect is one of the most well-established features of contemporary urbanization, but its social significance and implications for human health, environmental justice, and urban infrastructure remain important frontiers [7]. In the case of drought and flooding, climate change projections include an increased frequency of extreme precipitation events interspersed by longer, more extreme drought events [8]. Drought is not only a concern of arid-climate cities [9,10]. In fact, the effects of drought as a limit to development and as a periodic threat to the functioning of mesic urban systems are becoming more prevalent [10]. However, drought in its various manifestations is often seen as a temporary pulse to be waited out, rather than as a stimulus for sustainable design, adaptation, and planning [11]. On the other hand, stormwater management is a strong stimulus for sustainable design, adaptation, and planning in all cities [12,13]. Flood risk is a similar stimulus for river cities, as is the inevitable risk of sea level rise and ocean flooding in coastal cities. All are examples of extreme conditions that are expected to intensify in the future. They often have disproportionate impacts on the most vulnerable populations, such as youth and the elderly, low-income people, people of color, and immigrants [14].

Urban design has been defined by and associated with architects, landscape architects, and planning professionals since the mid-20th century. However, as cities worldwide must adapt to economic restructuring, mass migrations, and climate change, urban design is quickly evolving to include civil society, community actors, engineers, and city managers [15,16]. The urban design process, as it is more broadly conceived, has the capacity to incorporate and give physical form to our growing scientific understanding of urban regions as social-ecological systems [17] facing new climate change risks and vulnerabilities. Pre-industrial cities were largely designed to provide human comfort in sometimes-hostile conditions. Later, modern infrastructure put distance between urban design and direct climate response, but recent innovations in climate-appropriate architecture and design are beginning to address this dichotomy. We argue that an integrated design-ecology nexus has the potential to stimulate a new era of sustainable urban development based in novel systems of deliberative decision-making and governance. The rationale for this integration was well articulated by Tanner *et al.* [16], and here we both expand on that rationale and present a model to make it happen. This merging of design and ecology will advance urban sustainability most quickly when a nexus of feedbacks between them is purposefully created to promote adaptive change and risk mitigation in both urban neighborhoods and regions. As our species continues to become more urban [18], this transformational design-ecology nexus is increasingly critical. It will help existing and new cities meet the challenges of growing in sustainable ways while becoming more resilient to severe, climate-induced events. In essence, society cannot adequately meet these challenges if urban design and ecology continue to be practiced as separate fields and in isolation from residents, city governance, and decision-making.

In this paper we: (1) summarize the relationships among design, infrastructure, and urban development with particular emphasis on the importance of joining the three to achieve urban climate resilience and enhance sustainability; (2) discuss the need for urban ecology to move from an ecology *of* cities [19,20] to an ecology *for* cities based on a knowledge-to-action agenda; (3) present a model for a transformational urban design-ecology nexus, and; (4) demonstrate the efficacy of our model with several case studies.

DESIGN, INFRASTRUCTURE, AND SUSTAINABLE URBAN DEVELOPMENT

Urban design became a professional practice in the middle of the 20th century as cities around the world faced the enormous challenges of reconstruction after World War II [21]. Through its development, the large-scale *tabula rasa* or "blank slate" approach to modern urban planning was replaced by code-based district or patch-scale urban design. This shift emphasized reconstructing or reconnecting to traditional cities and their fine-scale streets, blocks, and open space rather than the large-scale infrastructural emphasis of modern city planning. The discipline of landscape ecology helped shape regional planning approaches to, for example, greenfield development [22,23] and more recent movements labeled landscape or ecological urbanism have focused on post-industrial brownfield redevelopment [24,25]. However a more comprehensive and integrative design-ecology nexus is needed to consider greenfield, brownfield, and greyfield areas of cities while operating at scales of existing and new neighborhood development [26].

It is well known that urbanization is increasing globally, and converting ever-greater areas of land worldwide. In addition to direct land conversion, urban processes, lifestyles, and investments are transforming existing cities while also influencing lands far distant from traditionally recognized cities [27,28]. Thus, an urban perspective on and concern for sustainability is increasingly relevant to a large proportion of the Earth's land covers, ecosystems, and populations. Many cities are already addressing the need to advance urban sustainability with formal sustainability plans. However, these plans often focus on particular sectors, infrastructure types, or processes, such as transportation, water, or energy [29]. In many cases these sustainability plans are not adequate to prepare cities for the crises and tipping points that so many cities and new urban areas face currently or will experience in the future [3,4]. These plans often suffer from a lack of integration and holistic vision that connects across sectors and allows a networked approach to planning for the future. Many sustainability plans also tend to have a narrow focus on existing "hard" infrastructures (e.g., transportation, water supply, and sanitary and stormwater treatment systems) and on "low hanging fruit" green infrastructures (e.g., parks and public trees [30,31,32]). Furthermore, these plans have rarely been integrated into the smaller-scale neighborhood approach of contemporary urban design practice [33,34].

Urban infrastructure takes many forms. Engineered infrastructure, or "gray" infrastructure, is the capital investments that move or house people, goods, water, waste, and energy and the associated guidelines for their construction, operation, maintenance, and rehabilitation. In most contemporary cities, what is in place today was largely deployed over the past century during the era of the "Sanitary City" [34], when less was known about the environmental, social, and climate change impacts of the physical design of systems. Gray infrastructure was built to protect against environmental hazards by, in general, overdesigning or designing to control the environmental factors [35]. This reduced the likelihood that infrastructure would fail—a "fail safe" goal. This kind of engineered resilience is rigid and unyielding; it is not adaptive and imparts large inertias on urban systems [3]. New approaches that couple infrastructure and urban ecological services are needed, as evidence accumulates that these twentieth century infrastructure design principles often make urban systems more vulnerable to extreme climate events [16,36,37].

Urban infrastructure that incorporates natural elements is often called "green" infrastructure. This includes traditional examples, such a parks and street trees, but also includes more novel urban components such as community gardens and multi-purpose and multi-function stormwater management facilities. Virtually all cities are also characterized by a variety of water features that provide a range of ecosystem services, including rivers and streams, lakes, and fountains. Collectively these are known as "blue" infrastructure. Because green and blue infrastructure features take advantage of natural structures and ecological processes, they are surprisingly adaptable to a changing future, and thus impart resilience to urban systems far more than do inertia-bound gray infrastructures. In fact, there are numerous options for using green and/ or blue infrastructure designs instead of gray infrastructure [38]. We argue, as did Tanner *et al.* [16], that the best strategy for moving in this direction is for the urban design process to expand beyond the realm of engineers and planners in order to include urban ecologists, landscape architects, and deliberative decision-making and governance. Representatives of all disciplines and perspectives should be a part of the design process from beginning to end [39].

Urban design has social impacts across a range of social-ecological scales and can contribute to integrating the ecological functioning of green and blue infrastructure with the finer scales of individuals, urban neighborhoods, and communities. At the level of individuals, environmental stewardship can promote relaxation, mitigate stress, create self-confidence, and strengthen self-sufficiency. At the collective level of neighborhoods and communities, environmental stewardship can help establish trust, strengthen social cohesion, and enhance knowledge sharing [40,41,42]. Investments in urban design, in concert with green, blue, and turquoise infrastructure (Text Box 1), have economic impacts in the real estate sector and the workforce. For example, numerous studies have demonstrated the positive impacts of proximity to parks, community gardens, and tree-lined streets on residential real estate value [43,44] and on commercial activity in business districts [45]. As cities initiate sustainability plans and move towards more green, blue, and turquoise infrastructure designs, society will need a larger urban natural infrastructure management workforce. Many cities have embarked on widespread tree planting and green infrastructure initiatives, with the need for ongoing maintenance by professionals [46,47,48,49]. With a growing interest in new green, blue, and turquoise infrastructure, urban environmental restoration and maintenance jobs offer an opportunity for those with limited formal education or specialized job training to establish what are broadly referred to as "green-collar" careers. Furthermore, engaging in environmental restoration work may provide benefits to workers beyond simple employment, including exposure to and interactions with nature [50], which is a limited "commodity" in many urban locations. In short, cities are habitat for people, so the urban design process should include city residents and integrate a social component into design objectives and actions.

> Green and blue infrastructure are familiar concepts and features of cities. These two terms, and color designations, delineate services provided by terrestrial versus aquatic urban ecological features and ecosystems. But these two infrastructure designations may not cover all urban ecological features and associated services. For example, urban wetlands are an important component of "natural" infrastructure in virtually all cities. While some urban wetlands are found in, or have been rehabilitated to be in, somewhat natural states, many of these wetland systems were designed and constructed to provide specific services. These services include wastewater treatment, stormwater management, recreation and aesthetics, and habitat. Wetlands are found where terrestrial and aquatic ecosystems meet, and are considered ecotone ecosystems. They have ecological functions and structures that are both terrestrial and aquatic in character [51]. As such, urban wetlands are both green and blue infrastructure. When one combines green and blue, the result is the color turquoise. Thus, we posit that urban wetlands are unique enough in the urban matrix to warrant the term "turquoise infrastructure".

Box 1: The Colors of Urban Design: Turquoise Infrastructure.

MOVING FROM AN ECOLOGY *OF* CITIES TO AN ECOLOGY *FOR* CITIES

Calls for improved integration of ecological knowledge with the theory and practice of urban design are decades old [52,53], including those from the UNESCO Man and the Biosphere Program (e.g., [54,55]). Yet successful integration is rare and is concentrated in a few institutions and opportunities [56,57,58]. We argue that there is a pressing need to institutionalize these interdisciplinary connections and to extend their scope. Sustaining both human well-being and the integrity of the biosphere continues to be a test of public decision-making and institutions, and a major challenge to the academy and its linkages to civic processes. The scope of this challenge is daunting because of the complexity of sustainability, whose three pillars—environment, society, and economy—encompass many dimensions, including social values, ethics, equity, justice, connectedness and heritage, ecological integrity, and economic viability. Our argument for and approach to a transformational design-ecology nexus is also a call for us to move from an ecology *of* cities, as described by Pickett *et al.* [19] and Grimm *et al.* [20], to an ecology *for* cities where urban ecologists, designers, planners, engineers, residents, and others are actively pursuing a more sustainable future (per [3]).

If sustainability is a socially-generated process, then social-ecological resilience is its mechanism [3]. Resilience as a theory highlights the need for understanding how complex adaptive systems develop and evolve, the role of disturbance and disruption, the drivers of collapse of complex social-ecological systems, the nature of reorganization after collapse, and the various modes of adaptation that permit reorganization [59,60,61]. However, standard resilience theory has largely neglected an explicit consideration of design, and both its intended and unintended consequences, and instead has focused on the more comfortable realm of policy and management. Thus, we need an improved understanding of urban sustainability that extends beyond the founding concerns of resilience theory with natural resources management to include the design process. The design realm often focuses on resilience as a socially- or politically-driven desire to build back after some disaster [62]. This is an important step, and additional analyses of resilience, such as those focused on transportation and associated urban form [63] exemplify additional steps. This transformative integration is required to achieve urban systems resilience, particularly in the face of extreme conditions, such as heat, drought, and floods. In the next section we describe a model for this transformation.

A MODEL FOR TRANSFORMING THE URBAN DESIGN-ECOLOGY NEXUS

Merely positing that a transformational change needs to occur is insufficient if positive outcomes are the goal. Thus, we present a model for how to advance an urban design-ecology nexus to meet the complex needs of cities facing a future of change and uncertainty. We propose that any model needs to have certain performance features. It needs to be flexible so that ecological boundaries—as spaces of action—can change and diversify, mirroring everyday environmental realities. It needs to be adaptive, inclusive, and responsive to our increasingly inequitable societies and human migrations. And it needs to be future thinking in ways that value diverse forms of imagination and communication—for example, drawing has a powerful role in capturing a shared sense of future possibilities [64]. These features of an urban design-ecology approach will substantially improve the capacity of urbanized jurisdictions and urban regions to improve their sustainability plans and actions. Our focus on extreme events, exemplified primarily by heat waves, drought, and floods, turns the attention of the design and policy-making communities to conditions that are not always present, but that nonetheless are important for making sustainable urban designs flexible and adaptable. Because design addresses many spatial scales, myriad clients, and is situated in a diversity of land covers and neighborhoods, engaging ecological science will vastly enhance the effectiveness of both the design process and of urban ecology [65].

First and foremost, our model is inclusive and involves all requisite parties in the urban design process (Figure 1B). As we noted in the section above, this transdisciplinary design process should include urban social and ecological scientists, landscape architects, city residents, and students of all related disciplines in addition to professional architects, engineers, and planners. A key contrast between our model and the status quo (Figure 1A) is that our model of the design process is non-linear and includes critical feedbacks, throughout the process, among key stakeholders, urban ecological research, and knowledge derived from previous projects [39]. In terms of sustainability science, this is the co-production or co-generation of urban design [66] that is manifest by providing a collaborative environment for urban designs to be envisioned and implemented [67,68].

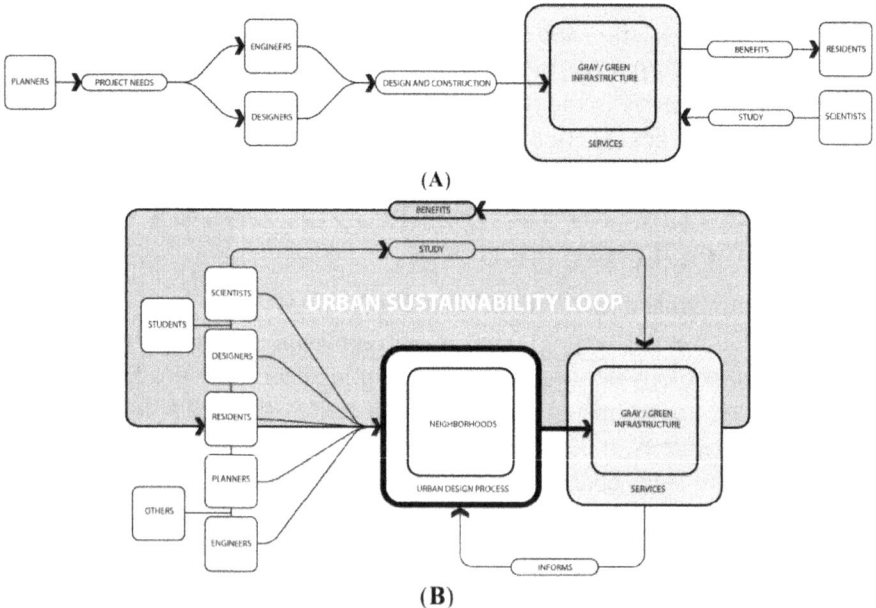

Figure 1: (A) (Top): The typical way in which urban infrastructure is conceived, de-

signed, constructed, and evaluated. Focus is often on only one or a few services and the process is linear; (B) (Bottom): The integrated approach, proposed here, for implementing urban infrastructure in which the design is co-produced in a transdisciplinary way, and the infrastructure is designed to provide multiple services. This approach is not linear; rather, it has several key feedback loops that we show as contained in an "urban sustainability loop"; this loop is highlighted in yellow for emphasis. Note that "Scientists" includes urban social and ecological scientists and "Others" may include elected officials, business leaders, famous individuals, powerful interest groups, *etc.*

Cities are designed to be human habitat, and as such *Homo sapiens* are the "ecosystem engineers" [69] of urban systems. No attempt to create a new, transformative, integrated approach to urban design can be successful without fully considering the social structures and dynamics of the city [57], including those that are value-laden and normative [70]. Paying attention to the well-being of populations that have been historically overburdened by environmental hazards and excluded from policy and planning decisions is also crucial in the design of sustainable and equitable cities and regions [54]. For this reason equity is a key tenet of sustainability [1,71,72]. A transformative urban design process must incorporate justice and equity [73], and strategies for doing so should be informed by history, including socio-spatial analyses of patterns of settlement, displacement, and segregation of vulnerable populations, and the policy and planning procedures that shaped those patterns [74].

Urban governance is defined as a mode of public decision-making that involves public, private, and non-profit actors at all levels of city management [75]. Governance theory can be used as an organizing framework to examine and understand macro-scale shifts from top-down state-centered approaches towards more interactive policy-making processes and hybrid arrangements that involve diverse interactions among multiple social actors [76,77,78]. In the context of urban environmental governance, a shift towards multi-sector activities can be a response to climate change uncertainty and a management strategy for enabling robust urban ecosystem services as well as systemic adaptation and resilience [79,80,81].

While urban planners and policy-makers have long addressed environmental issues, the constellation of groups associated with the governance of urban ecological processes has shifted in recent decades from hierarchical to polycentric structures [82,83,84,85,86]. In many cities, urban environmental stewardship groups have become an essential component of governance that regulates ecosystem services [87]. These groups include organizations that work to conserve, manage, monitor, advocate for, or educate about a range of quality-of-life issues regarding urban resources. They include community development groups focused on environmental issues and environmental groups focused on community-based quality-of-life issues [88]. Taken together, these constellations of civic groups can form a large component of novel urban governance systems and can be important stakeholders in our transformational model for the urban design process [89].

Designers can choose to locate their work within democratic politics in a range of modes. The consensus mode, as an ideal, is sometimes not realizable. In addition, it has been criticized for being hegemonic when there is no possibility of final reconciliation. In these situations designers can work in an agonistic mode of democratic politics by visualizing that which is repressed and destroyed by the consensus of post-political democracy [90], and expanding capacities for collective production and self-governance [89].

As we noted earlier, our model for a transformative urban design-ecology nexus is future-oriented and anticipatory. For this reason, a critical component of our model is creative visioning. Transformation does not happen instantaneously. Thus, any evolution of the urban design process must involve the education and training of a cadre of: (1) ecologically literate urban designers and engineers; (2) design-literate, engineering-conscious ecologists; (3) broad-thinking and holistically inclined planners, and; (4) place-aware and activist city residents. There are myriad ways to accomplish these educational goals [33,90], using both place-based and virtual approaches. Interdisciplinary design studios, workshops, community charrettes, and teaching are all critical, and may be facilitated by Internet-based interactions. This may be particularly valuable for facilitating interactions with government or non-profit employees who may find it difficult to travel to face-to-face studios or charrettes. Although direct interactions may be quite valuable, virtual interactions can also facilitate the contributions of community members in design processes that bring together experts from distant locations who might not otherwise be able to participate. Furthermore, virtual interactions can be digitally archived and used again, thus allowing activities to be virtual in both space and time. Our model does not intend to be prescriptive about the specifics, as long as transformative goals, inclusiveness, and a focus on the future are central to the approach. One might imagine that place-based "transformative design studios" would be located at academic institutions or at larger design firms, with strong connections to local planners and community members. A more ambitious approach would network and integrate a number of these inclusive, place-based studios across cities, with virtual connections to each other and to planning and to decision-making entities in their respective cities.

Regardless of the approach, it is critical that these "transformative design studios" or "arenas of inclusive design" produce: (1) methodologies for rigorously assessing the effectiveness of ecologically-informed designs both on the "drawing board" and in the field; (2) platforms for the transdisciplinary sharing of sustainable ecological design insights; (3) lasting institutional and organizational collaborations, preferably across a spectrum of climatic conditions for the inter-city networks, to maintain these efforts; and (4) an expansion of general theories of inclusive urbanism [17,26,91,92]. This integrated network of students, community members and leaders, graduates, scientists, and teachers should connect across the nation, and ultimately the globe. But at the same time they must engage their local institutions and government agencies that have a history of acting in isolation despite the growing calls for sustainable urban design, revitalization, and regional integration. In short, our ambition for a transformational urban design-ecology nexus has the ultimate goal of changing the way cities are designed, managed, and inhabited based on a deeper social, ecological, technological, and political understanding of urban sustainability and of the urban experience.

EXAMPLES OF CHALLENGES AND SUCCESSES

Fine-scale experimental urban design provides a unique opportunity to explore transformative new approaches to the design process [93,94]. In our inclusive model for transforming the urban design process (Figure 1), all players involved in design would also be continually involved in these on-the-ground design initiatives or experiments. As such, our model includes both the co-production of design and the co-production of knowledge through design experimentation, all happening in a safe space—real or virtual—for collaboration and communication. Below we briefly describe several case studies in which this co-production approach has been attempted or used successfully.

- Goodyear AZ USA: This project was focused on an experimental streetscape project that was initiated at the behest of city planners from the City of Goodyear, AZ, USA. A team of urban ecologists, landscape architects and students, and social scientists was assembled to design and implement four streetscape options with differing landscaping, water use, and microclimate impacts. The streetscape was to be built on a 25 m × 750 m plot of vacant land along Goodyear's main boulevard, near City Hall (Figure 2). City planners and community members from an adjacent historic neighborhood provided ideas on and local knowledge that informed designs drafted by an upper division landscape architecture studio class at Arizona State University. Prior to construction of the experimental streetscape, data were to be collected on human-scale microclimate to quantify physical characteristics that relate to people's perceptions of their immediate climate (Figure 3).

Educational signage was planned to keep the neighborhood and users of this unique streetscape park updated on new findings and project progress, and residents were to be periodically be surveyed about their project-related perceptions and values. After the project was constructed, biophysical and social monitoring were to continue to quantitatively identify the best, and most desirable, streetscape designs, as well as the success of the park as an education and outreach tool. This project should have been a showcase example of how integrating ecology, design, social science, and policy moves us from an ecology *of* cities to an ecology *for* cities (per [3]). Unfortunately, the entire project was ultimately cancelled after more than a year of negotiations between City officials, the landowner, and lawyers. The park was to be located on the edge of a remediated Superfund site, and the legal challenges associated with this finally made the project untenable. As this paper was published, City officials were looking for another site for the park. This is an example of the challenges that can be faced when trying to implement our transdisciplinary ecology-design nexus model.

Figure 2: Street view of the property in Goodyear, AZ, USA where the experimental demonstration streetscape will be constructed. The park was to be on the left and the historical neighborhood is on the extreme right (photo credit: D. Childers).

Figure 3: Micrometeorological towers used to monitor variables related to people's

perceptions of their immediate climate. Towers include sensors that measure air temperature, relative humidity, wind speed, and light at 2 m height, as well as soil temperature, soil moisture, and irrigation water use (photo credit, B. Ruddell).

- Baltimore MD USA: Over ten years, academic urban design studios have been instrumental in building a dialogue about sustainable urban development between scientific research by the Baltimore Ecosystem Study Long-Term Ecological Research Program (BES LTER) and local non-profit and government organizations. The urban design-ecology nexus developed by the BES LTER Urban Design Working Group (UDWG) connects the neighborhood scale of urban design practice with the ecological theory of patch dynamics. Urban neighborhoods in the Baltimore region are distinct land-cover neighborhood patches within larger watershed and human ecosystem frameworks. Early studios organized by the BES UDWG focused on connecting the emerging social infrastructure of the Gwynns Falls trail system to new urban land cover and hydrological data [95]. This work, based on an open urban stream network, continued in the underground piped Watershed 263 in West Baltimore [96]. It also included proposals for the redevelopment of the Middle Branch waterfront with the office of Baltimore City Planning. More recently design studios with the Baltimore Office of Sustainability and two neighborhood groups—The Harlem Park Neighborhood Association/Parks and People and the Historic East Baltimore Action Coalition—provided examples of how ecosystem science can be integrated with local grassroots efforts to rejuvenate neighborhoods suffering from disinvestment. Working with vacant lots and unused city space, social capital and job generation were a critical part of these green and blue infrastructure proposals. Examples of the resulting designs can be found at [97,98]
- The Urban Complex, NJ USA: Between 2002 and 2010, the urban design and landscape architecture firm Till Design undertook two client-based, professional-practice, brownfield redevelopment projects. The first, Monroe Center for the Arts, includes an art community in two old factory buildings and two new buildings surrounded by a plaza on two blocks located on a former wetland on the western edge of Hoboken, NJ. The second, Celadon, is a new high-rise, high-density, transit-oriented, mixed-use development located on a municipal dump on Newark Bay, in Elizabeth NJ. In both projects, urban ecology informed the design in three ways to increase the economic, cultural, and environmental value of the projects:
  - • The inclusion of green infrastructure such as stormwater features, roof

gardens, vegetated plazas, and waterfront promenades.

- The inclusion of a phased approach where micro-scale social-ecological engagement was supported from the very start of the project. This allowed the project to debut publicly, with elements such as a drawing club on a parking lot, a temporary film garden, a low-tech kayak launch, and a sound game. Other elements included fountains, roof gardens accessible by public elevators, the co-production of a "bay music" sound signature, and a ferry to Manhattan.
- The inclusion of innovative financing in short, medium, and long concurrent cycles that were designed and launched to allow both of these environmentally activist real-estate development to happen [99,100].
- Sacramento, CA, USA: The Building Healthy Communities site in Sacramento, CA has embarked upon a process of community planning in order to engage residents and develop comprehensive and inclusive neighborhood action plans that inform advocacy. Building Healthy Communities is a program of the California Endowment, a private, state-wide health organization that aims to improve the health status of all Californians by considering what happens outside of the doctor's office and beyond people's individual choices. In 2010, the California Endowment initiated a 10-year initiative called Building Healthy Communities (BHC). It chose 14 cities for its initial application based on the health status and challenges to health faced by the residents of these cities and on the potential to inspire broader policy changes that could create healthier environments. The Sacramento BHC consists of nearly 20 small neighborhoods spanning both incorporated and unincorporated areas of South Sacramento. More than 65,000 residents call the Sacramento BHC home. While these residents experience high rates of economic, educational, and health challenges, they also are represented by engaged and supportive elective officials. The residents are served by many active and collaborative grassroots advocacy organizations working to address these disparities. In 2014, a collection of organizations, led by the non-profit organization Ubuntu Green, held a series of events that culminated in a week-long charrette, engaging residents to develop community action plans around several issues including access to healthy food, economic development and housing, pedestrian and bike safety, and environmental health. Residents could participate in person or by responding to Internet-based surveys and commenting on social media outlets. All materials were presented in three languages and events ran through the day and into the evening, providing as much time as possible for participation. The resulting

community action plans will be released in early 2015 and will be used to inform advocacy for the organizations and resident groups associated with the BHC.

SUMMARY

Urban development takes a wide variety of forms around the world, from the older, traditional compact city, to the bustling metropolis of the 19th and 20th century, to the low density car-based exurb, to the informal megacity. Cities of all forms would benefit from the transformative nexus of ecology and design that we propose here in order to advance climate change resilience and enhance future sustainability. This design-ecology nexus does not prescribe a narrow range of solutions for green design or urban sustainability; rather, it is a transdisciplinary process of reiteration, feedbacks, deliberation, and debate between social and natural systems. An ecology *for* cities is a call for action-based ecological research and knowledge that is part of a new urban design process working at all scales of urban decision-making, from individual households to neighborhoods to regions. We argue that this inclusive, creative process will produce new and innovative solutions that will allow tomorrow's cities to be better prepared for a climate-uncertain future.

ACKNOWLEDGMENTS

The authors of this paper are members of the Urban Sustainability Research Coordination Network that is supported by the U.S. National Science Foundation (NSF) through Grant No. 1140070. Additional support has been provided by the NSF to DLC through the CAP LTER Program (Grant No. 1026865), to MLC, JMG, BM, and STAP through the BES LTER Program (Grant No. 1027188), and to MLC through Grant No. 0844778.

AUTHOR CONTRIBUTIONS

All authors contributed substantially to the ideas, concepts, and work presented in this paper. All authors were also involved in the preparation of the manuscript and have approved the submitted form.

REFERENCES

1. Gibson, R.B. Beyond the pillars: Sustainability assessment as a framework for effective integration of social, economic, and ecological considerations in significant decision-making. *J. Environ. Assess. Policy Manag.* 2006, *8*, 259–280.

2. Leach, M.; Scoones, I.; Stirling, A. *Dynamic Sustainabilities: Technology,*

Environment, Social Justice; Earthscan: New York, NY, USA, 2010.
3. Childers, D.L.; Pickett, S.T.A.; Grove, J.M.; Ogden, L.; Whitmer, A. Advancing urban sustainability theory and action: Challenges and opportunities. *Landsc. Urban Plan.* 2014, *125*, 320–328.
4. Pickett, S.T.A.; Boone, C.G.; McGrath, B.P.; Cadenasso, M.L.; Childers, D.L.; Ogden, L.A.; McHale, M.; Grove, J.M. Ecological science and transformation to the sustainable city. *Cities* 2013, *32*, S10–S20.
5. Papaneck, V. *Design for the Real World: Human Ecology and Social Change*; Academy Chicago Publishers: Chicago, IL, USA, 2005.
6. Thackara, J. *In the Bubble: Designing for a Complex World*; The MIT Press: Cambridge, UK, 2006.
7. Huang, G.L.; Zhou, W.Q.; Cadenasso, M.L. Is everyone hot in the city? Spatial pattern of land surface temperatures, land cover and neighborhood socioeconomic characteristics in Baltimore, MD. *J. Environ. Manag.* 2011, *92*, 1753–1759.
8. Milly, P.C.D.; Betancourt, J.; Falkenmark, M.; Hirsch, R.M.; Kundzewicz, Z.W.; Letternaier, D.P.; Stouffer, R.J. Stationarity is dead: Whither water management. *Science* 2008, *319*, 573–574.
9. Platt, R.H.; Rowntree, R.A.; Muick, P.C. Sustainability in urban ecosystems: Beyond an object of study. In *The Ecological City: Preserving and Restoring Urban Biodiversity*; University of Massachusetts-Amherst Press: Amherst, MA, USA, 1994; pp. 49–65.
10. Hunt, D.V.L.; Rogers, C.D.F. A Benchmarking System for Domestic Water Use. *Sustainability* 2014, *6*, 2993–3018.
11. Tingstad, A.H.; Groves, D.G.; Lempert, R.J. Paleoclimate Scenarios to Inform Decision Making in Water Resource Management: Example from Southern California's Inland Empire. *J. Water Res. Plan. Manag.* 2014, *140*, 04014025.
12. Desimini, J. Civic space in regional frameworks: Resilient approaches to urban design. In *Resilience in Ecology and Urban Design: Linking Theory and Practice for Sustainable Cities*; Pickett, S.T.A., Cadenasso, M.L., McGrath, B., Eds.; Springer: New York, NY, USA, 2013; pp. 307–318.
13. Shannon, K. Eco-engineering for water: From soft to hard and back. In *Resilience in Ecology and Urban Design: Linking Theory and Practice for Sustainable Cities*; Pickett, S.T.A., Cadenasso, M.L., McGrath, B., Eds.; Springer: New York, NY, USA, 2013; pp. 163–182.
14. Adaman, F. Power inequalities in explaining the link between natural

hazards and unnatural disasters. *Dev. Chang.*2012, *43*, 395–407.
15. McGrath, B. *Urban Design Ecologies*; John Wiley & Sons: London, UK, 2012.
16. Tanner, C.J.; Adler, F.R.; Grimm, N.B.; Groffman, P.M.; Levin, S.A.; Munshi-South, J.; Pataki, D.E.; Pavao-Zucherman, M.; Wilson, W.G. Urban ecology: Advancing science and society. *Front. Ecol. Environ.* 2014, *21*, 574–581.
17. Pickett, S.T.A.; Cadenasso, M.L.; McGrath, B. *Resilience in Ecology and Urban Design: Linking Thoery and Practice for Sustainable Cities*; Springer: New York, NY, USA, 2013; p. 499.
18. United Nations. *World Urbanization Prospects: The 2011 Revision Highlights*; United Nations: New York, NY, USA, 2012; p. 50.
19. Pickett, S.T.A.; Burch, W., Jr.; Dalton, S.; Foresman, T.W.; Rowntree, R. A conceptual framework for the study of human ecosystems in urban areas. *Urban Ecosyst.* 1997, *1*, 185–199.
20. Grimm, N.B.; Grove, J.M.; Pickett, S.T.A.; Redman, C.L. Integrated approaches to long-term studies of urban ecological systems. *BioScience* 2000, *50*, 571–584.
21. Shane, D.G. *Urban Design since 1945: A Global Perspective*; John Wiley & Sons: London, UK, 2011.
22. McHarg, I. *Design with Nature*; The Natural History Press: New York, NY, USA, 1969.
23. Forman, R.T. *Land Mosaics: The Ecology of Landscapes and Cities*; Cambridge University Press: Cambridge, UK, 1995.
24. Waldheim, C. *Landscape Urbanism*; Princeton Architectural Press: New York, NY, USA, 2006.
25. Mostafavi, M.; Doherty, G. *Ecological Urbanism*; Lars Muller: Zürich, Switzerland, 2010.
26. Spirn, A.W. Ecological Urbanism: A Framework for the Design of Resilient Cities. Available online: http://www.annewhistonspirn.com/pdf/Spirn-EcoUrbanism-2012.pdf (accessed on 1 November 2012).
27. Seto, K.C.; Reenberg, A.; Boone, C.G.; Fragkias, M.; Haase, D.; Langanke, T.; Marcotullio, P.; Munroe, D.K.; Olah, B.; Simon, D. Urban land teleconnections and sustainability. *Proc. Natl. Acad. Sci. USA* 2012, *109*, 7687–7692.
28. Boone, C.G.; Redman, C.L.; Blanco, H.; Haase, D.; Koch, J.; Lwasa, S.; Nagendra, H.; Pauleit, S.; Pickett, S.T.A.; Seto, K.C.; *et al.* Reconceptualizing land for sustainable urbanity. In *Rethinking Urban*

Land Use in a Global Era; Seto, K.C., Reenberg, A., Eds.; MIT Press: Cambridge, UK, 2014; pp. 313–330.

29. Newman, P.; Jenkins, I. *Cities As Sustainable Ecosystems: Principles and Practices*; Island Press: Washington, DC, USA, 2008.

30. Larson, E.K.; Earl, S.; Hagen, E.M.; Hale, R.; Hartnett, H.; McCracken, M.K.; McHale, M.; Grimm, N.B. Beyond restoration and into design: Hydrological alterations in aridland cities. In *Resilience in Ecology and Urban Design: Linking Theory and Practice for Sustainable Cities*; Pickett, S.T.A., Cadenasso, M.L., McGrath, B., Eds.; Springer: New York, NY, USA, 2013; pp. 183–210.

31. Novotny, V. Water-energy nexus: Retrofitting urban areas to achieve zero pollution. *Build. Res. Inf.* 2013, *45*, 489–604.

32. Sze, J.; Gambirrazio, G. Eco-cities without ecology: Constructing ideologies, valuing nature. In *Resilience in Ecology and Urban Design: Linking Theory and Practice for Sustainable Cities*; Pickett, S.T.A., Cadenasso, M.L., McGrath, B., Eds.; Springer: New York, NY, USA, 2013; pp. 289–297.

33. Bargmann, J. Just ground: A social infrastructure for urban landscape regeneration. In *Resilience in Ecology and Urban Design: Linking Theory and Practice for Sustainable Cities*; Pickett, S.T.A., Cadenasso, M.L., McGrath, B., Eds.; Springer: New York, NY, USA, 2013; pp. 347–354.

34. Melosi, M.V. *The sanitary city: Urban Infrastructure in America from Colonial Times to the Present*; Johns Hopkins University Press: Baltimore, MD, USA, 2000; p. 600.

35. Ahern, J. From fail-safe to safe-to-fail: Sustainability and resilience in the new urban world. *Landsc. Urban Plan.* 2011, *100*, 341–343.

36. Fiksel, J. Sustainability and resilience: Toward a systems approach. *Sustain. Sci. Pract. Policy* 2006, *2*, 14–21.

37. Anderies, J.M.; Janssen, M.A. Robustness of social-ecological systems: Implications for public policy. *Policy Stud. J.* 2013, *41*, 513–536.

38. Beatley, T. *Green urbanism: Learning from European cities*; Island Press: Washington, DC, USA, 2000; p. 491.

39. Felson, A.J. The design process as a framework for collaboration between ecologists and designers. In *Resilience in Ecology and Urban Design: Linking Theory and Practice for Sustainable Cities*; Pickett, S.T.A., Cadenasso, M.L., McGrath, B., Eds.; Springer: New York, NY, USA, 2013; pp. 365–382.

40. Glover, T.D.; Shinew, K.J.; Parry, D.C. Association, sociability, and

civic culture: The democratic effect of community gardening. *Leis. Sci.* 2005, *27*, 75–92.

41. Svendsen, E. Cultivating resilience: Urban stewardship as a means to improving health and well-being. *Restorative Commons: Creating Health and Well-Being through Urban Landscapes*; Campbell, L., Wiesen, A., Eds.; General Technical Report. U.S. Department of Agriculture, Forest Service, Northern Research Station: Newtown Square, PA, USA, 2009; pp. 58–87.

42. Peters, K.; Elands, B.; Buijs, A. Social interactions in urban parks: Stimulating social cohesion? *Urban For. Urban Green.* 2010, *9*, 93–100.

43. Morales, D.J. The contribution of trees to residential property value. *J. Arbor.* 1980, *6*, 305–308.

44. Anderson, L.M.; Cordell, H.K. Influence of trees on residential property values in Athens, Georgia (USA): A survey based on actual sales prices. *Land. Urban Plan.* 1988, *15*, 153–164.

45. Wolf, K. Business district streetscapes, Trees and consumer response. *J. For.* 2005, *103*, 396–400.

46. Baltimore Center for Green Careers. Available online: http://www.baltimoregreencareers.org (accessed on 12 March 2015).

47. City of Los Angeles. Million Trees LA. Available online: http://www.milliontreesla.org/mtabout1.htm (accessed on 12 March 2015).

48. City of Seattle. City of Seattle Comprehensive Plan. Available online: http://www.seattle.gov/dpd/Planning/Seattle_s_Comprehensive_Plan/ComprehensivePlan/ (accessed on 12 March 2012).

49. City of Phoenix. Tree and Shade Master Plan. Available online: https://www.phoenix.gov/parkssite/Documents/T%20and%20A%202010.pdf (accessed on 12 March 2015).

50. Falxa-Raymond, N.; Svendsen, E.; Campbell, L.K. From job training to green jobs: A case study for a young adult employment program centered on environmental restoration in New York City, USA. *Urban For. Urban Green.* 2013, *12*, 287–295.

51. Mitsch, W.J.; Gosselink, J.G. *Wetlands: Third Edition*; John Wiley and Sons: New York, NY, USA, 2000.

52. Spirn, A.W. *The Granite Garden: Urban Nature and Human Design*; Basic Books: New York, NY, USA, 1984.

53. Steiner, F. Urban landscape perspectives. *Landsc. Urban Plan.* 2014, *3*, 342–350.

54. Di Castri, F.; Baker, F.; Hadley, M. The Man and the Biosphere Program

as an Evolving System. *Ambio* 1981, *10*, 52–57.

55. Boyden, S.; Millar, S.; Newcombe, K.; O'Neill, B. *The Ecology of a City and Its People: The Case of Hong Kong*; Australian National University Press: Canberra, Australia, 1981.

56. Pickett, S.T.A.; McGrath, B.; Cadenasso, M.L. The ecology of the metacity: Shaping the dynamic; patchy, networked, and adaptive cities of the future. In *Resilience in Ecology and Urban Design: Linking Theory and Practice for Sustainable Cities*; Pickett, S.T.A., Cadenasso, M.L., McGrath, B., Eds.; Springer: New York, NY, USA, 2013; pp. 463–489.

57. Boone, C.G. Social dynamics and sustainable urban design. In *Resilience in Ecology and Urban Design: Linking Theory and Practice for Sustainable Cities*; Pickett, S.T.A., Cadenasso, M.L., McGrath, B., Eds.; Springer: New York, NY, USA, 2013; pp. 47–61.

58. Grove, J.M. Ecological and social linkages in urban design projects: A synthesis. In *Resilience in Ecology and Urban Design: Linking Theory and Practice for Sustainable Cities*; Pickett, S.T.A., Cadenasso, M.L., McGrath, B., Eds.; Springer: New York, NY, USA, 2013; pp. 355–360.

59. Holling, C.S. Cross-scale morphology, geometry, and dynamics of ecosystems. *Ecol. Monogr.* 1992, *62*, 447–502.

60. Wu, J.; Wu, T. Ecological resilience as a foundation for urban design and sustainability. In *Resilience in Ecology and Urban Design: Linking Theory and Practice for Sustainable Cities*; Pickett, S.T.A., Cadenasso, M.L., McGrath, B., Eds.; Springer: New York, NY, USA, 2013; pp. 211–229.

61. Pickett, S.T.A.; McGrath, B.; Cadenasso, M.L.; Felson, A.J. Ecological resilience and resilient cities. *Build. Res. Inf.* 2014, *42*, 143–157.

62. Vale, L.J.; Campanella, T.J. *The Resilient City: How Modern Cities Recover from Disaster*; Oxford University Press: New York, NY, USA, 2005; p. 376.

63. Newman, P.; Beatley, T.; Boyer, H. *Resilient Cities: Responding to Peak Oil and Climate Change*; Island Press: Washington, DC, USA, 2009.

64. Marshall, V. Ecological Urban Design Visuality and Landscape. *Nakhara J. Environ. Des. Plan.* 2014, *10*, 113–126.

65. Felson, A.J.; Bradford, M.A.; Terway, T.M. Promoting Earth Stewardship through urban design experiments. *Front. Ecol. Environ.* 2013, *11*, 362–367.

66. Schellnhuber, H.J.; Crutzen, P.J.; Clark, W.C.; Hunt, J. Earth system analysis for sustainability. *Sci. Policy Sustain. Dev.* 2005, *47*, 11–25.

67. Lemos, M.C.; Morehouse, B.J. The co-production of science and policy in integrated assessments. *Glob. Environ. Chang.* 2005, *15*, 57–68.
68. Robinson, J.; Tansey, J. Co-production, emergent properties and strong interactive social research: The Georgia Basin Futures Project. *Sci. Public Policy* 2006, *33*, 151–160.
69. Lawton, J.H.; Jones, C.G. Linking species and ecosystems: Organisms as ecosystem engineers. In *Linking Species and Ecosystems*; Jones, C.G., Lawton, J.H., Eds.; Chapman and Hall: New York, NY, USA, 1995; pp. 141–150.
70. Rozzi, R.; Pickett, S.T.A.; Palmer, C.; Callicott, J.B. *Linking Ecology and Ethics for a Changing World: Values, Philosophy, and Action*; Springer: New York, NY, USA, 2014.
71. Agyeman, J.; Evans, T. Toward just sustainability in urban communities: Building equity rights with sustainable solutions. *Ann. Am. Acad. Polit. Soc. Sci.* 2003, *590*, 35–53.
72. Agyeman, J. *Sustainable Communities and the Challenge of Environmental Justice*; NYU Press: New York, NY, USA, 2005.
73. Pickett, S.T.A.; Buckley, G.L.; Kaushal, S.S.; Williams, Y. Social-ecological science in the humane metropolis. *Urban Ecosyst.* 2011, *14*, 319–339.
74. Lord, C.H.; Norquist, K. Cities as Emergent Systems: Race as a Rule in Organized Complexity. *Environ. Law* 2010, *40*, 551–597.
75. Stoker, G. Governance as theory: Five propositions. *Int. Soc. Sci. J.* 1998, *50*, 17–28.
76. Rhodes, R.A.W. The new governance: Governing without government. *Polit. Stud.* 1996, *44*, 652–667.
77. Jordan, A. The governance of sustainable development: Taking stock and looking forwards. *Environ. Plan. C Govern. Policy* 2008, *26*, 17–33.
78. Fisher, D.; Svendsen, E. Hybrid arrangements within the environmental state. In *Routledge International Handbook of Social and Environmental Change*; Lockie, S., Sonnenfeld, D., Fisher, D., Eds.; Routledge: London, UK, 2014; pp. 179–189.
79. Bulkeley, H.; Betsill, M. *Cities and Climate Change: Urban Sustainability and Global Environmental Governance*; Routledge: New York, NY, USA, 2003.
80. Adger, W.N.; Arnell, N.W.; Tompkins, E.L. Successful adaptation to climate change across scales. *Glob. Environ. Chang.* 2005, *15*, 77–86.
81. Daily, G.C.; Polasky, S.; Goldstein, J.; Kareiva, P.M.; Mooney, H.A.;

Pejchar, L.; Ricketts, T.H.; Salzman, J.; Shallenberger, R. Ecosystem services in decision making: Time to deliver. *Front. Ecol. Environ.* 2009, *7*, 21–28.

82. Weber, E.P. A new vanguard for the environment: Grass-roots ecosystem management as a new environmental movement. *Soc. Natl. Resour.* 2000, *13*, 237–259.

83. Kempton, W.; Holland, D.C.; Bunting-Howarth, K.; Hannan, E.C. Local environmental groups: A systematic enumeration in two geographical areas. *Rural Sociol.* 2001, *66*, 557–578.

84. Sirianni, C.; Friedland, L. *Civic Innovation in America: Community Empowerment, Public Policy, and the Movement for Civic Renewal*; University of California Press: Berkeley, CA, USA, 2001.

85. Andrews, K.T.; Edwards, B. The organizational structure of local environmentalism. *Mobiliz. Int. Quart.* 2005, *10*, 213–234.

86. Svendsen, E.S.; Campbell, L.K. *The Urban Ecology Collaborative Assessment: Understanding the Structure, Function, and Network of Local Environmental Stewardship*; U.S.D.A. Forest Service Northern Research Station: New York, NY, USA, 2005.

87. Ernstson, H.; Sorlin, S.; Elmqvist, T. Social movements and ecosystem services—The role of social network structure in protecting and managing urban green areas in Stockholm. *Ecol. Soc.* 2008, *13*, Article 39.

88. Fisher, D.R.; Campbell, L.K.; Svendsen, E.S. The organizational structure of urban environmental stewardship.*Environ. Polit.* 2012, *12*, 26–48.

89. Svendsen, E. Storyline and design: How civic stewardship shapes urban design in New York City. In *Resilience in Ecology and Urban Design: Linking Theory and Practice for Sustainable Cities*; Pickett, S.T.A., Cadenasso, M.L., McGrath, B., Eds.; Springer: New York, NY, USA, 2013; pp. 269–287.

90. Mouffe, C. Art and Democracy: Art as an Agnostic Intervention in Public Space. *Open Art Public Issue* 2008, *14*, 6–15.

91. Hardt, M.; Negri, A. *Commonwealth*; Harvard University Press: Cambridge, UK, 2009.

92. Miss, M. Remixing messages: A call for collaboration between artists and scientists. In *Resilience in Ecology and Urban Design: Linking Theory and Practice for Sustainable Cities*; Pickett, S.T.A., Cadenasso, M.L., McGrath, B., Eds.; Springer: New York, NY, USA, 2013.

93. Ellin, N. Integral urbanism: A context for urban design. In *Resilience in Ecology and Urban Design: Linking Theory and Practice for Sustainable*

Cities; Pickett, S.T.A., Cadenasso, M.L., McGrath, B., Eds.; Springer: New York, NY, USA, 2013; pp. 63–78.

94. North, A.; Waldheim, C. Landscape urbanism: A North American perspective. In *Resilience in Ecology and Urban Design: Linking Theory and Practice for Sustainable Cities*; Pickett, S.T.A., Cadenasso, M.L., McGrath, B., Eds.; Springer: New York, NY, USA, 2013.

95. Felson, A.J.; Pickett, S.T.A. Designed experiments: New approaches to studying urban ecosystems. *Front. Ecol. Environ.* 2005, *3*, 549–556.

96. Hager, G.W.; Belt, K.T.; Stack, W.; Burgess, K.; Grove, J.M.; Caplan, B.; Hardcastle, M.; Shelley, D.; Pickett, S.T.A.; Groffman, P.M. Socioecological revitalization of an urban watershed. *Front. Ecol. Environ.* 2013, *11*, 28–36.

97. Harlem Park, Baltimore, example. Available online: http://sce.parsons.edu/studiocourses/BaltimoreStudio2010/(accessed on 12 March 2015).

98. Baltimore Ecosystem Study Urban Design Working Group. Available online: http://beslter.org/frame4-page_3i_07.html (accessed on 12 March 2015).

99. Felson, A.J.; Oldfield, E.E.; Bradford, M.A. Involving ecologists in shaping large-scale green infrastructure projects. *BioScience* 2013, *63*, 882–890.

100. McGrath, B.; Marshall, V.; Cadenasso, M.L.; Grove, J.M.; Pickett, S.T.A.; Plunz, R.; Towers, J. *Designing Patch Dynamics*; Columbia University: New York, NY, USA, 2007.

Chapter 9

INDUSTRIAL ECOLOGY AND ENVIRONMENTAL LEAN MANAGEMENT: LIGHTS AND SHADOWS

Giuseppe Ioppolo[1], Stefano Cucurachi[2], Roberta Salomone[1], Giuseppe Saija[1] and Luigi Ciraolo[1]

[1]Department of Economics, Business, Environment and Quantitative Methods (SEAM), University of Study of Messina, P.zza Puglatti 1, Messina 98122, Italy

[2]Department CML-Industrial Ecology, Leiden University, 2300 RA Leiden, The Netherlands

ABSTRACT

Current industrial production is driven by increasing globalization, which has led to a steady increase in production volumes and complexity of products aimed at the pursuit of meeting the needs of customers. In this context, one of the main tools in the management of customer value is Lean Manufacturing or Production, though it is considered primarily as a set of tools to reduce the total cost of the resources needed to achieve such needs. This philosophy has recently been enriched in the literature with case studies that link Lean Management (LM) with the improvement of environmental sustainability. The consequence is an expansion of the Computer Integrated Manufacturing (CIM); indeed, CIM, currently, combining and integrating the key business functions (e.g., business, engineering, manufacturing, and information management) with a view of the life cycle, does not highlight the strategic role of the environmental aspects. In order to deal with the increasingly rapid environmental degradation that is reflected in society, in terms of both economy and quality of life, Industrial Ecology (IE) introduced a new paradigm of principles and instruments of analysis and decision support (e.g., Life Cycle Assessment—LCA, Social Life Cycle Assessment -SLCA, Material Flow Account—MFA, *etc.*) that can be considered as the main basis for integrating the environmental aspects in each strategy, design, production, final product, and end of life management, through the re-engineering of processes and activities towards the development of an eco-industrial system. This paper presents the preliminary observations based on a analysis of both theories (LM-IE) and provides a possible assessment of the key factors relevant to their integration in a "lean environmental management", highlighting both positives (lights) and possible barriers (shadows).

INTRODUCTION

In March 2010, the European Commission presented its strategy "Europe 2020", and took note that the Western model of economic development has triggered a process of environmental degradation of difficult resolution. Such problem overlaps, both with the problem of limited natural resources, and also with the inequality of resource availability across geographies. Europe 2020 is a decennary strategy aiming at a structural transformation of the European economy in order to overcome the economic crisis and the challenges of the next decade through competitive and sustainable forms [1]. To achieve this, the Europe 2020 Strategy addresses the issue of growth, declining in three main areas [1]:

(1) Smart growth (promotion of knowledge, innovation, education, and digital society);

(2) Sustainable growth (production more efficient in terms of resources and greater competitiveness);

(3) Inclusive growth (more jobs, skills, and combating poverty).

In this way, the Europe 2020 Strategy identifies the eco-industries (also defined environmental industries or companies environmental technology oriented) as the enabled actors in order to do grow wealth and employment without causing serious damage to the environment [1]. For example, the strategy promotes those industries that produce so-called *"enabling technologies"*. These kinds of industries allow to greatly increase the performance and ability of the user without increasing the consumption of resources (as in the case of information technology); moreover, this industrial model allows to reach the famous "decoupling" between economic growth and resource use, including the creation of wealth and environmental impacts.

The priority initiatives indicated in Europe 2020, were further underlined by the European Commission in its Communication "Rio+20: Towards the green economy and better governance" [2]. Indeed, the EU highlights the combination of competitiveness and the green economy, outlining the strategic directions (e.g., shifting the aim on R&D, introducing clean technologies, pointing to an industrial policy for green growth as a means to improve competitiveness, creating new jobs working to achieve a low-carbon and efficient use of resources). The increasing attention to environmental aspects, measured along the entire life cycle of a product/process, either by the company or by all possible stakeholders, needs to align the production model to that of environmental management, in order to create a new business model that is green-oriented. The authors, in order to deal the increasingly rapid environmental degradation, in terms of both economy and quality of life, answering to the Europe 2020 strategy, carry out several key factors useful to the re-engineering of processes and activities towards the development of an eco-industrial system.

A possible way is furnished by the Industrial Ecology (IE) theory, that introduced a new paradigm of principles and instruments of analysis and decision support (e.g., LCA, SLCA, MFA, *etc.*); these tools can be considered as the main basis for assessing and integrating the environmental aspects in each strategy, design, production, final product, and end of life management.

This paper presents the preliminary considerations based on an examination of two accredited theories (also defined strategies and philosophies):

(1) The Lean Management (LM) as a highly competitive production model [3,4];
(2) The Industrial Ecology (IE) as a framework of principles and tools of environmental analysis.

The aim of this paper is to highlight the key factors relevant to their integration in a environmental lean management system, both of positive terms (lights) and possible barrier (shadows).

THEORY AND METHODS-LEAN MANAGEMENT AND INDUSTRIAL ECOLOGY

The Lean Production (LP) is based on the principles and processes introduced in the Toyota Production System (TPS), and was defined as "doing more with less" [5,6,7]. Womack, formalizing the principles of this theory, characterizes it as a system of measures and methods that, in a holistic approach, have the potential to reduce the production factors. It follows that the lean production model is reflected in the degree of competitiveness of the entire business system that adopts it. The LP is therefore a strategy or philosophy that promotes the use of practices, such as the *kanban*, a type of scheduling system, the total quality management (TQM) and just-in-time (JIT), to minimize scrap/waste and improve the performance of a company [8,9]. The LP initially broken down into four areas/strategic phases of production: Product development, supply chain, also called Kanban supermarket [10,11,12], the management of the workshop, and after-sales service. Thanks to a continued and applied research the LP has become a vital model for the entire business system, the Lean Management (LM) [13], emphasizing the expansion of the management production management of the organization in all its activities. In this way, the LM is often seen as a set of tools that compete for the reduction of the overall cost, and at the same time, are intended to improve the quality of manufactured products.

From an operational perspective, the LM is achieved through the adoption of a business model of integrated production or CIM (Computer Integrated Manufacturing), which represents the most complete form of integration between the different areas of a production system in an automated factory (*i.e.*, design, engineering, production, quality control, production planning, and marketing).

The CIM can be considered as a standard for the industry and may be automated through the use of Information Communication Technology (ICT), the LM is performed throughout the business life cycle (from design Computer Aided Design-Computer Aided Engineering-Finite Element Analysis (CAD-CAE-FEA, to the stages of production Computer Aided Manufacturing (CAM), to sales and after-sales service, a unique business model for optimal management of resources, Enterprise Resource Planning (ERP)).

Only recently the scientific literature proposes studies linking the philosophy of the LM with the improvement of environmental sustainability [14,15,16,17,18]. These studies suggest that lean production is more than a set of lean tools to optimize production efficiencies; it is a *modus operandi* and a mindset that must be integrated into production systems, in a systematic way, in order to achieve sustainability [16,17]. Obviously, the goal is to guide the LM towards a green perspective, and involves the crossing of conceptual limits of the same LM. In fact, although certainly the LM ensures better operating results, such as lower inventory through a lean warehouse, higher quality in all business processes, and overall shorter timescale due to a complete synchronization between events, however, it does not internalize the environmental perspective in their principles [6,10].

The above considerations are reflected in the "new inventory paradigm" introduced by Chikan [19], from which one can clearly understand what the relationships to other processes and functions within enterprises geared to profitability are. The paradigm does not focus on environmental issues, but on a number of mediating factors between lean production and financial performance, which is the real driving force in the spread of this theory. The paradigm can be detailed as follows:

(1)   Establish a strategic vision: The *Value* must be defined jointly for each

product family, along with a cost-based target on the value perceived by the customer;

(2) Identify and establish teams (identify the flow): *Value Stream*, namely the monitoring and identification of responsibility in all activities that are necessary specifications from the design, management of orders and deliveries, launch, production and final delivery to the customer;

(3) Identify the products (slide the flow): *Flow*, rethink specific work practices and tools to eliminate returns, scrap and arrests (of any kind), so that the design, order and production of a specific product may proceed in a continuous manner;

(4) Identify processes: *Pull*, flow only active when pulled to the next step;

(5) Review the layout of the factory: *Perfection*, the complete elimination of muda (waste) so that all the activities cascade contribute to the creation of value;

(6) Select an appropriate strategy *Kanban* (pull-push adaptive approach);

(7) Improve continuously, while maintaining the excellent results.

In combination with the above principles, companies can apply different types of environmental practices to improve their productivity in the use of natural resources, such as energy and materials, and to reduce the related environmental impacts of their activities [4].

The report "Lean and Green", presented by Zokaei *et al.* [20], and the studies carried out by Glavic and Lukman [21] and Lozano [22] provide an overview of some of these practices (e.g., the use of cleaner forms of production, the introduction of models production-oriented eco-efficiency). For Cagno *et al.* [23], cleaner production is an initiative of environmental protection that the company puts in place with a view to prevention. This initiative is designed to minimize the amount of waste and emissions (negative output), while maximizing production (positive output).

Using the definition given by González del Río [24], clean technology can be assimilated to changes in production processes that reduce the amount of waste and pollution generated in the production process or during the entire life cycle of the product.

This approach directs the company towards a greater focus on the analysis and measurement of the flows of materials and energy produced by the relationship between the company itself and the reference area or crossing. In addition, the management, in a pro-active way, adopts strategies to reduce these flows in their industrial processes, using both the improvements in the management (for example, the integration of the quality management system with the environment management system-TQEM). Another strategy is the introduction of technological advances applied to production (introduction of the Environmental Technology Innovations (TEIs)), in order to develop an eco-friendly company policy (e.g., the sustainable issue are the waste reduction, the improvement in the integrated management of the water cycle, the reduction of greenhouse gas emissions and losses of warmth, the attention to other impacts, such as acoustic, *etc.*).

Korhone [25] rationalizes the above principles in seven critical success factors, which are necessary to achieve eco-efficiency in production:

(1) The reduction in the intensity of the material (de-materialization);
(2) The reduction in energy intensity (de-energized);
(3) The reduction of dispersion of toxic substances;
(4) The improvement the recyclability of materials introduced into the production process;
(5) The substitution of input materials with other resources more environmentally sustainable;
(6) The reduction in the degree of persistence of the pollutants;
(7) The value increasing of the intangible component of each product.

These success factors are combined with the need to incorporate eco-efficiency already in the vision and business strategy, along with a growing availability of cleaner technology, which transforms the green market in the new field on which to compete. In contrast, the process of industrialization, compared with an ever-increasing demand following the enlargement of markets, is linked to the follow main factors of environmental impact:

(1) The high level of air pollution caused by combustion processes;
(2) The pollution of water bodies, determined by discharges of process waste to which are added the discharges of household;
(3) The production and accumulation of waste, increasingly complex, and difficult to recycle;
(4) The production of new materials, chemical products, such as plastics and synthetic products are not biodegradable.

In this way, another important role is played by the community of all stakeholders, which have to be involved in the decision process in order to build a shared knowledge and a collective consciousness changing its behaviors [26].

As a result of the foregoing, the world of scientific research and technological innovation, has set the goal of finding the appropriate solutions, creating new technologies applied to production cycles, making it possible to prevent and/ or reduce pollution and to reduce, to minimal amount, the substances emitted (output) and the natural resources used (input).

At the same time, the increased level of awareness on the environmental risks has spread an ever-increasing attention on environmental aspects along the whole life cycle of a product/process; in this sense, it has reinforced the need to assess and share information, such as [27]:

(1) The environmental damage caused by the technological processes implemented and processed products;

(2) The actions necessary for modernization or modification of the technological process;

(3) The results of the comparative analysis for the definition of alternative technologies that lead to the production of the same product or a product change, reducing environmental impacts.

In response to these pressures from the legislature, the marketplace and the community, and the local, national, and international levels that, among the various theories have assumed a significant role in the Industrial Ecology (IE). The central concept and characterizing for the IE is an industrial system that should not be considered isolated from its surrounding systems, but in a position of continuous exchange. The IE deals with the systematic study of patterns of industry in relation to the natural and social systems involved, and it is designed to optimize such trade in terms of sustainability [28,29].

Tibbs [30] and Ayres and Ayres [15] summarize the concept of industrial ecology and translate it into an "industrial ecosystem", stressing the importance of thinking about how the characteristics of the natural ecosystem can be translated into an industrial ecosystem. The introduction of the concept of time-scale completes this concept, which in the anthropic system (techno-sphere) is relevant and not unlimited, allowing the analysis of the interactions that take place between all parties (competitive interactions and/or cooperation). Focusing on the principles of industrial ecology, they could be considered as the main support for the introduction of environmental considerations in all activities, from strategy to design, to production, to the realization of the product throughout the life cycle, through the king-engineering of processes and activities in an eco-industrial system.

To achieve these goals, it is necessary to understand the factors influenced a sustainable society; these are the scarcity of resources, the need for materials and the growth of energy consumption, the approach to the de-materialization and substitutability for sustainable development. In this sense, there are three major transitions [31,32]:

(1) The transition from fossil fuels to renewable energy sources;
(2) The transition from linear flows of an economy closed-loop material that gives economic value to the new secondary raw materials, with a view to re-use and recycling;
(3) The transition from the exploitation of nature and biodiversity to its protection.

The IE is a paradigm for environmental management principles and makes use of tools for environmental analysis and the definitions of compatible choices. It can then be finalized by the interpretation of the transformation of the industrial system connected to the load capacity of the territorial system in which it is rooted. In this sense, the IE has been regarded as a broad holistic framework to guide the transformation of the industrial system towards a model of sustainable production.

The profound change that is involved in the management of production from a linear model into a closed-loop model closely resembles the cyclical flows of ecosystems, drawing on the biological concept of ecology, which is the branch of biology that deals with the study of relationships that occur between organisms and the physical environment that hosts them [15].

Therefore, the IE seeks to structure the industrial models in a substantially closed loop, to benefit both economically and environmentally, not focusing on individual industrial processes, but proposing a new sustainable economy, based on a clear understanding of the interaction between the world of production and environmental system.

For sighting this result you need to avoid that the analyses are partial and simplistic, ignoring important variables and, above all, leading to unintended consequences.

The IE, in fact, addresses the entire lifecycle of a product-process, focusing on the use of resources and materials in relation to the analysis of energy flows, and through modeling systems, investigates the impact-relations environment, using a multidisciplinary and interdisciplinary approach, which aims to suggest options and more sustainable choices.

The IE uses a set of tools oriented to the product, e.g., the analysis of the life cycle (LCA), which seeks to ensure that, in the examination of an industrial process or product, all its interactions and impacts on the environment are fully accounted, from the extraction of raw materials, the processes of production, use and disposal of the product; this tool has been further specialized with the assessment of costs throughout the life cycle (LCC) and the evaluation of social aspects (SLCA), which are also particularly important in the industrial process in terms of human resource management [33].

Together with the LCA, there are a number of other tools oriented to the study of relationships between environmental systems (e.g., enterprise/manufacturing district and geographical areas of production and sales). Among these services, one may include the analysis of the flow of materials (MFA), which adopts a macroscopic approach, the analysis of the flow of a substance (SFA) with a microscopic approach, tools based on input-output tables that use statistical data and are most suitable for studying environmental reports at national/international regulations (e.g., physical input-output table (PIOT), and the ecological network analysis (ENA)).

For the analysis of the flows of materials and energy the first law of thermodynamics on the conservation of matter is applied, allowing a consistent and comprehensive collection of input and output flows, and stocks within the study area, *i.e.*, the system under observation [34].

The analysis of the flows of materials and energy can be used at both the global and local scales. Globally, this analysis of flows can help determine the extent to which human activities are influencing/impacting the natural systems of the Earth (e.g., hydrological cycle, the carbon cycle, and the nitrogen cycle). At the local level or enterprise level, these instruments introduce research methods that calculate the mass balances of industrial processes, the results of which can be used to ensure that all resources are fully valued in a sustainable manner.

The IE also introduces tools that are particularly suited to the realization of environmental information that is clear and easily understandable, and especially consumer-oriented. For example, the concept of ecological footprint has been developed by Ayres [35] in response to the debate on the concept of "carrying capacity". The carrying capacity is defined as "the population of a given species that can be sustained indefinitely in a given habitat without permanently damaging the ecosystem on which it depends" [36].

Thus, the ecological footprint expresses the theoretical area (in a single indicator that is "global hectare") used by man to produce biological resources it consumes and to absorb the waste it generates (including CO_2 resulting from its use of energy) [37].

Finally, again, based on the LCA, it is appropriate to highlight the Design for Environment (also called eco-design), which is a widely used approach for the improvement of environmental performance. The Design for Environment integrates environmental considerations throughout the life cycle of the product from the earliest stages of product design [38].

What has been said up to now shows the intrinsic connections between the LM and IE approaches. Based on the previous discussion of both theories, it is possible to identify a set of common evaluation criteria, which use the same data to measure the relationship between the company (techno-sphere) and the environment (eco-sphere).

The full integration of the two theories could represent the real success factor by single operating unit up to complex industrial system (e.g., District), observing such phenomena from the local to global/international level (Figure 1).

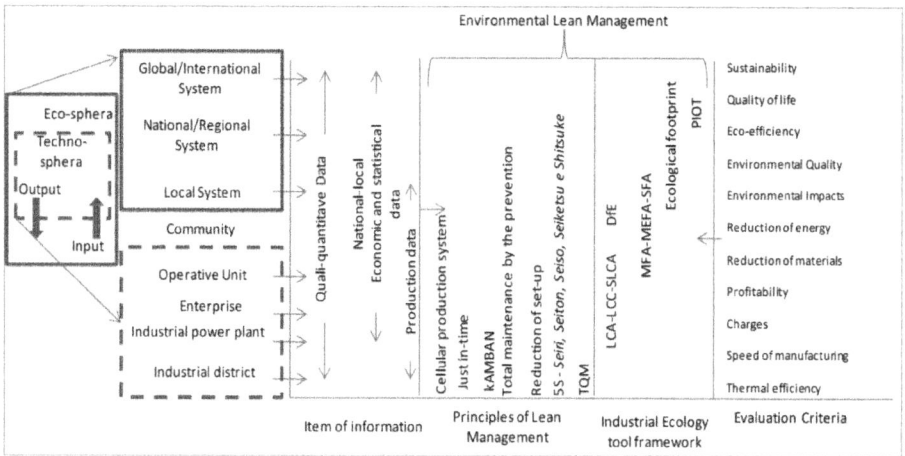

Figure 1: Selection of LM principles and IE tool, on the basis of common evaluation criteria and the same data types in order to support an environmental lean management model.

In this context, a lean company, although not necessarily green-oriented (rather generally, companies must make trade-offs among multiple objectives that are not fully compatible), has time for conversion and response to new market trends, including environmental, quicker and, therefore, is particularly ready to adopt a model of environmental lean management.

RESULTS

Environmental lean management fully integrates with the green-oriented LM model, combining the basic principles of LM (*i.e.*, the five basic principles [39]) with those green principles (e.g., better use of natural resources and the reduction environmental impact [4]) in order to create a unique integrated model.

The environmental model of LM introduces the environmental variable along all processes, imposing eco-efficiency in production and use of resources, materials, and energy, and also introducing the goals of reducing environmental impacts, and of sharing the environmental information for environmental awareness spread along the flow/value chain.

From integration of LM and CT strengths and opportunities, is possible to carry out the common targets' need to push the launch of the Environmental lean management system (Figure 2).

The environmental model of lean management is focused on overcoming the traditional forms of savings, which include reduction of overproduction, waiting time, transport optimization, adequacy in the answers to the problems, sizing inventory flow management, and reduction of defects [40].

This idea is reflected, for example, in the study by Moreira *et al.* [4], which explores other forms of loss in value (environmental impacts, energy consumption, material consumption, and emissions) to demonstrate that environmental liabilities are hidden inside of classic lean manufacturing.

From what has been stated in the previous paragraph, it is possible declare that a lean environment acts as a catalyst to facilitate environmental sustainability.

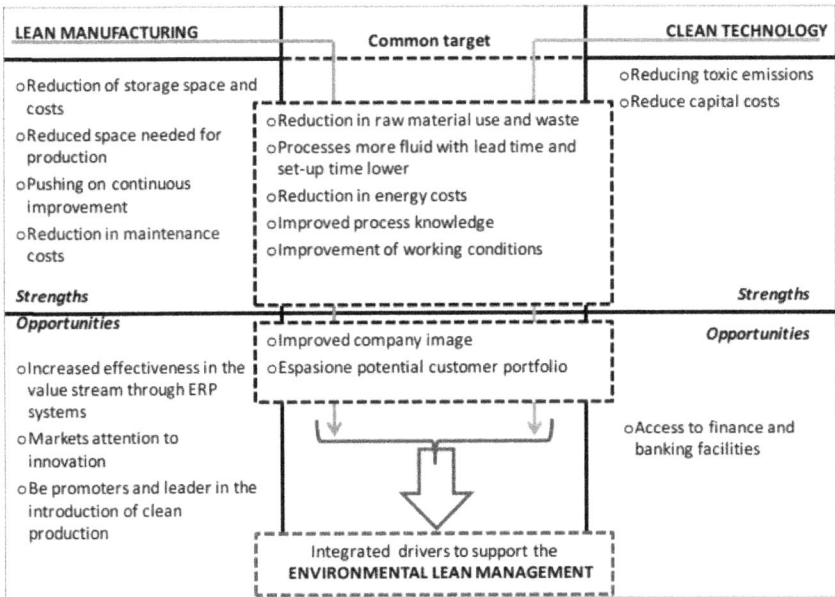

Figure 2: Synthesis of the main strengths and opportunities of LM and CT, in order to support the environmental lean management model.

Combining the LM aims [40] with the needs of sustainability that society asks [41], is possible to identify the vision of the environmental model of LM, that is to realize an efficient production system, effective and sustainable, within the social and economic aspects connected to the environmental and territorial conditions, from a long value chain perspective (from local community to global market).

In this perspective, the IE provides, through its principles, and its analysis and decision support tools, an effective response to transit from one model to that of environmental lean management.

In fact, the IE, in terms of support for the environmental assessment, applied to the entire corporate system allows:

(1) An overall assessment of the environmental risks arising from technological process and finished products;
(2) The quantitative analysis of environmental burdens related to the flows of materials and energy used in the process, as well as the flow of output products (In-put = output + stock);
(3) The evaluation of the recovery of secondary raw materials derived from waste and recycling processes that should be taken into account;
(4) The measurement of the consumption of all the raw materials and the use of recovered energy in the process;
(5) An identification of any other industrial waste streams secondary (solid, liquid, and gaseous fuels) to be taken into account (hidden flows);
(6) An identification of any flows of co-products harmful to the environment.

Then, IE could be considered as a paradigm of tools that environmental lean management adopts to maximize resource productivity and close the cycle of movement of resources within the techno-sphere, preserving the limited natural resources and minimizing the amount of production waste and related emissions.

This model can be used as a basis for determining the prospects and the direction of change necessary for a sustainable society, increasing the productivity of resources through the recirculation of what, before, was "lost" or "released" to the ecosphere.

A company that intends to reconvert according to a model of environmental lean management must start from the identification of the main sources of waste, and proceed with the elimination of the arising inefficiencies, and, then, develop more sustainable products, addressing green markets.

With the introduction of IE in the *modus operandi* of the LM, it is possible to define a new set of objectives that integrate environmental aspects within the financial-productive concerns:

(1) Reduce material and energy use for a product, including services in the course of his life;
(2) Reduce emissions, dispersion and the creation of toxic substances during his lifetime;
(3) Increase the amount of recyclable materials;
(4) Maximize sustainable use of renewable resources;
(5) Minimize the intensity of service for products and services;
(6) Extending the useful life of a product;
(7) Assess and minimize the environmental impact during the life of the product;
(8) Have a "functional economy" is a way to replace the products with services;
(9) Use "reverse logistics", which means that all efforts are used in order to reuse products and materials;
(10) Increase the efficiency of a product during use.

The final result is that both theories-philosophies have a strategic function and totally present, if one (IE) is applied in the other (LM).

DISCUSSIONS OF POSITIVE ASPECTS AND POSSIBLE BARRIERS

The correlation between LM and IE, for the study of new business models, leans green-oriented, points to a number of involved factors. The most important pressures for companies of a certain size, are the standards and rules set by customers (particularly international customers) that require environmental safeguards, in the perspectives of both cost savings and increased profits for shareholders; moreover, the increasing of communication and sharing of policies and performance, in terms of action and responsible environmental protection, with all the stakeholders-community, represents another relevant matter.

Instead, for SMEs, the most important drivers are represented by the target of cost savings. In addition, for such companies, environmental sustainability remains a consequence dictated by the legal obligations that continue to play a decisive role (mandatory system under "command and control").

Availability and accessibility to technology are two other key factors. However, an ever-increasing level of technological development is not proportional and uniform access to it. Limitations on physical-geographical, political-regulated, financial, and historical barriers are also slowing down in the field of technological innovation, the introduction of cleaner technology.

The three dimensions of sustainability (3P-society (people), environment (planet), economics (profit)), should, therefore, be extended by a fourth: The technological dimension [42,43,44]. The European Commission, in this sense, plays an important role as a catalyst for the development of European Technology Platforms (ETP) that promote the TEIs. In doing so, they gather key stakeholders operating in sectors with a high content of technology, innovation, and research, and sectors for which competitiveness and innovation, economic development and social development, depend on technological advances implemented in the medium and long term.

ETPs represent a model of technological governance, in which the legislature is a partner of the voluntary initiative of a European industrial system, with "bottom up" and "technology driven" approaches.

The ETP also try to overcome some of the main obstacles to the development of a lean green-oriented model from the point of view of industrial operators, namely the high costs of implementation of TEIs and the lack of tradition/ skills.

In this sense, the ETP represent a positive boost thanks to the mobilization of significant human and financial resources and can help ensure investment in research and development to improve the competitiveness of European companies and bring benefits to European citizens. Another significant aspect, not related to the company size, is represented by the objectives posed by the highly diffused Total Quality Management models. These are an ideal base for expansion and integration with other management models (e.g., environmental, social and ethical, safety, *etc.*), supporting a culture of environmental protection, through shared responsibility between the high direction (top-down) and all the parts inside and outside the company (bottom-up), and a culture of training extended to the entire lifecycle of the product, from the supplier to the end customer (life cycle management).

From an operational perspective, the literature shows that most of the efforts are focused, for instance, on product design and on the introduction of technology on single production step, but not along the entire supply chain; this strategy acts is more oriented on the management of end-life, rather than on an integrated production system. In this sense, the biggest problem is the mentality of the producers, who believe that the main role of industrial systems is to transform inputs into outputs, cost-effectively, using technologies and efficient processes, but does not consider it essential that sustainability can play a key role in competitiveness. This can also be attributed to the lack of widespread knowledge about the forces and mechanisms that support the implementation of practices for sustainable production are only also connected to a non-adequate notice of the improvements achieved by some green-oriented producers.

With regard to instruments introduced by the IE, there are strong resistances due to, e.g., the need for an investment in terms of money (external advice or preparation of a team within the company), or in terms of time and commitment to the collection of data, which are often not easily accessible and do not have adequate strength. For example, in the case of LCA, Udo de Haes [45] emphasizes the significant role to assess the complex interaction between a product and the environment, from cradle to cradle (as closed circle) and provide in-depth information on environmental impacts. LCA can be useful for manufacturing companies because it can demonstrate that the activities, processes, and materials used contribute to the production of large environmental impacts; these results can then serve as targets for improvement for green-oriented businesses.

However, as with other instruments, for the complexity of LCA studies can be particularly difficult for the entrepreneurs [46]. We have tried to overcome this negative aspect, which often prevents the application of a full LCA, through the application of a simplified version that also integrates into the paradigm the key features of lean management. Several other rigorous methodologies have been introduced to simplify LCA, supporting the adoption and dissemination [47].

Once all the difficulties that are typically associated with the application of IE are overcome, the analysis may prove its full ability to capture all environmental aspects. For example the DfE clearly points out the environmental impacts of a product that can be reduced more effectively, through the integration of environmental aspects in the early stages of the process (design) [48,49]. Environmental lean management can be a way to address the barriers and turn them into positive factors of competitiveness. In fact, this philosophy-strategy, if widely adopted, could represent a road map to guide long-term scientific and technological research, according to the needs of the industry, and to promote the application of technologies with a valid commercial return; environmental lean management can be considered an interdisciplinary model on which the industry may be interfaced with the research community and public authorities and regulators, in order to extend the dialog aimed at developing more effective and consistent standards and regulations, even in terms of promotion and development of adequate funding.

CONCLUSIONS

Improving the environmental performance in the manufacturing sector means decoupling economic performance from the environmental load of human activities. Environmental lean management represents a new frontier in terms of production efficiently, and is effectively able to reduce inputs of natural resources, materials and energy, or waste or pollutant outputs, while maintaining the assets of the techno-sphere separate from those of the ecosphere, in order to avoid environmental degradation.

The proposed integration could represent a best solution to achieve the sustainability. Current research in the areas of technologies for the production and management company systemic apparatus are increasingly linked to the productivity of resources, especially energy efficiency, and environmental assessment. Despite the growth in research on sustainable production, as well as efforts in the industrial sector, it is still difficult to find information on how to improve manufacturing operations and the cash flows of resources from the point of view of the producer [50].

The theme, still strongly present, meets a great deal of resistance among entrepreneurs, but it is a springboard for a future in the short term, in which the environmental variable has a weight comparable to or greater than the financial condition.

ACKNOWLEDGMENTS

We are particularly grateful to two anonymous referees for reading and for their comments.

AUTHOR CONTRIBUTIONS

These authors contributed fully and equally to this work. Giuseppe Ioppolo contributed to research design and coordinated with Luigi Ciraolo on the research project. Stefano Cucurachi analyzed the sources and literature. Roberta Salomone and Giuseppe Saija supervised the research project and carried out a detailed revision. All authors wrote the body of the paper and read and approved the final manuscript.

REFERENCES

1. Europa 2020. Available online: http://ec.europa.eu/europe2020/europe-2020-in-a-nutshell/flagship-initiatives/index_it.htm (accessed on 1 October 2013).
2. COM. Rio+20: Verso Un'economia Verde e una Migliore Governance. Available online: http://eur-lex.europa.eu/LexUriServ/LexUriServ.do?uri=COM:2011:0363:FIN:IT:PDF (accessed on 16 December 2013).

3. Gordon, P.J. *Lean and Greendprofit for Your Workplace and Environment*; Berrett-Koehler Publishers: San Francisco, CA, USA, 2001.
4. Moreira, F.; Alves, A.C.; Sousa, R.M. Towards eco-efficient lean production systems. *IFIP Adv. Inform. Commun. Technol.* 2010, *322*, 100–108.
5. Holweg, M. The genealogy of lean production. *J. Oper. Manag.* 2007, *25*, 420–437.
6. Womack, J.P.; Jones, D.T.; Roos, D. *The Machine that Changed the World*; Harper Perennial: New York, NY, USA, 1990.
7. Hopp, W.; Spearman, M.L. *Factory Physics*; McGraw-Hill/Irwin: New York, NY, USA, 2008.
8. Hofer, C.; Eroglu, C.; Hofer, A.R. The effect of lean production on financial performance: The mediating role of inventory leanness. *Int. J. Prod. Econ.* 2012, *138*, 242–253.
9. Guinipero, L.C.; Pillai, K.G.; Chapman, S.N.; Clark, R.A. A longitudinal examination of JIT purchasing practices. *Int. J. Logist. Manag.* 2005, *16*, 51–70.
10. Monden, Y. Adaptable Kanban system helps Toyota maintain just-in-time production. *Ind. Eng.* 1981, *13*, 29–46.
11. Ohno, T. *The Toyota Production System: Beyond Large-Scale Production*; Productivity Press: Portland, IN, USA, 1988.
12. Sugimori, Y.K.; Kusunoki, K.; Cho, F.; Uchikawa, S. Toyota Production System and Kanban System; materialization of just-intime and respect-for-human system. *Int. J. Prod. Res.* 1977, *15*, 553–564.
13. Warnecke, H.J.; Hiiser, M. Lean production. *Int. J. Prod. Econ.* 1995, *41*, 37–43.
14. Graedel, T.E. Industrial ecology: Definition and implementation. In *Industrial Ecology and Global Change*; Socolow, R., Andrews, C., Berkhout, F., Thomas, V., Eds.; Cambridge University Press: Cambridge, UK, 1994; pp. 23–41.
15. Ayres, R.U., Ayres, L.W., Eds.; *A Handbook of Industrial Ecology*; Edward Elgar Publishing: Northampton, MA, USA, 2002.
16. Robèrt, K.-H.; Schmidt-Bleek, B.; Aloisi de Larderel, J.; Basile, G.; Jansen, J.L.; Kuehr, R.; Thomas, P.P.; Suzuki, M.; Hawken, P.; Wackernagel, M. Strategic sustainable development: Selection, design and synergies of applied tools. *J. Clean. Prod.* 2002, *10*, 197–214.
17. Allwood, J.M. What is Sustainable Manufacturing? In *Sustainable Manufacturing Seminar Series*; Cambridge University: Cambridge, UK,

2005; pp. 1–32.

18. Seliger, G. *Sustainability in Manufacturing: Recovery of Resources in Product and Material Cycles*; Springer: Berlin, Germany, 2007.

19. Chikán, A. Managers' view of a new inventory paradigm. *Int. J. Prod. Econ.* 2011, *133*, 54–59.

20. Zokaei, K.; Martinez, F.; Vazquez, D.; Evans, B.; Sarmadi, K.; Peattie, L.; Pampanelli, A.; Hines, P.; Trivedi, N. *Best Practice Tools and Techniques for Carbon Reduction and Climate Change: Lean and Green Report*; Lean Enterprise Research Centre, Cardiff University: Cardiff, UK, 2010.

21. Glavic, P.; Lukman, R. Review of sustainability terms and their definitions. *J. Clean. Prod.* 2007, *15*, 1875–1885.

22. Lozano, R. Towards a better understanding of sustainability into companies' systems: An analysis of voluntary corporate initiatives. *J. Clean. Prod.* 2012, *25*, 14–26.

23. Cagno, E.; Trucco, P.; Tardini, L. Cleaner production and profitability: Analysis of 134 industrial pollution prevention (P2) project reports. *J. Clean. Prod.* 2005, *13*, 593–605.

24. Del Río González, P. Analysing the factors influencing clean technology adoption: A study of the Spanish pulp and paper industry. *Bus. Strateg. Environ.* 2005, *14*, 20–37.

25. Korhone, J. From material flow analysis to material flow management: Strategic sustainability management on a principle level. *J. Clean. Prod.* 2007, *15*, 1585–1595.

26. Ioppolo, G.; Saija, G.; Salomone, R. From coastal management to environmental management: The sustainable eco-tourism program for the mid-western coast of Sardinia (Italy). *Land Use Policy* 2013, *31*, 460–471.

27. Fijał, T. An environmental assessment method for cleaner production technologies. *J. Clean. Prod.* 2007, *15*, 914–919.

28. Ehrenfeld, J.R.; Gertler, N. Industrial ecology in practice: The evolution of interdependence at Kalundborg. *J. Ind. Ecol.* 1997, *1*, 67–79.

29. Graedel, T.E.; Howard-Grenville, J.A. *Greening the Industrial Facility: Perspectives, Approaches, and Tools*; Springer: New York, NY, USA, 2005.

30. Tibbs, H. Industrial Ecology. An Environmental Agenda for Industry. Global Business Network, 1993. Available online: http://www.hardintibbs.com/wp-content/uploads/2009/05/tibbs_indecology.pdf (accessed on 10 November 2013).

31. Kleijn, E.G.M. *Materials and Energy: A Story of Linkages*; Muller Visual: Leiden, The Netherlands, 2012.
32. Ioppolo, G.; Heijungs, R.; Cucurachi, S.; Salomone, R.; Kleijn, R. Urban Metabolism: Many open questions for future answers. In *Pathways to Environmental Sustainability: Methodologies and Experiences*; Salomone, R., Saija, G., Eds.; Springer International Publishing AG: Dordrecht, The Netherlands, 2014; pp. 23–32.
33. Zamagni, A. Life cycle sustainability assessment. *Int. J. Life Cycle Assess.* 2012, *17*, 373–376.
34. Hammer, M.; Giljum, S.; Bargigli, S.; Hinterberger, F. Material Flow Analysis on the Regional Level: Questions, Problems, Solutions. NEDS Working Paper, 2003. Available online: http://seri.at/wp-content/uploads/2009/09/Material-flow-analysis-on-the-regional-level1.pdf (accessed on 18 June 2014).
35. Ayres, R.U. Commentary in the utility of the ecological footprint concept. *Ecol. Econ.* 2000, *32*, 347–349.
36. Rees, W.E. Ecological footprints and appropriated carrying capacity: What urban economics leaves out. *Environ. Urban.* 1992, *4*, 121–130.
37. Wackernagel, M.; Lewan, L.; Hansson, C.B. Evaluating the use of natural capital with the ecological footprint: Applications in Sweden and Subregions. *Ambio* 1999, *28*, 604–612.
38. Allenby, B.R., Richards, D.J., Eds.; *The Greening of Industrial Ecosystems*; National Academy Press: Washington, DC, USA, 1994.
39. Womack, J.P.; Jones, D.T. *Lean Thinking*; Free Press: New York, NY, USA, 1998.
40. Bicheno, J.; Holweg, M. *The Lean Toolbox*; PICSIE Books: Buckingham, UK, 2000.
41. Lozano, R. Envisioning sustainability three-dimensionally. *J. Clean. Prod.* 2008, *16*, 1838–1846.
42. Elkington, J. *Cannibals with Forks: The Triple Bottom Line of 21st Century Business*; Capstone Publishing: Oxford, UK, 1997.
43. Lovins, A.B.; Lovins, L.H.; Hawken, P. A road map for natural capitalism. *Harv. Bus. Rev.* 1999, *77*, 145–158.
44. Jovane, F.; Yoshikawa, H.; Alting, L.; Boë, C.R.; Westkämper, E.; Williams, D.; Tseng, M.; Seliger, G.; Paci, A.M. The incoming global technological and industrial revolution towards competitive sustainable manufacturing. *CIRP Ann.-Manuf. Technol.* 2008, *57*, 641–659.
45. Udo de Haes, H.A. Applications of life cycle assessment: Expectations,

drawbacks and perspectives. *J. Clean. Prod.*1993, *1*, 131–137.

46. Ahlroth, S.; Nilsson, M.; Finnveden, G.; Hjelm, O.; Hochschorner, E. Weighting and valuation in selected environmental systems analysis tools e suggestions for further developments. *J. Clean. Prod.* 2011, *19*, 145–156.

47. Mori, Y.; Huppes, G.; Udo de Haes, H.A.; Otoma, S. Component manufacturing analysis: A simplified and rigorous LCI method for the manufacturing stage. *Int. J. Life Cycle Assess.* 2000, *5*, 327–334.

48. Keoleian, G.A.; Menerey, D. Sustainable development by design: Review of life cycle design and related approaches. *J. Air Waste Manag. Assoc.* 1994, *44*, 645–668.

49. Müller, K.G.; Court, A.W.; Besant, C.B. Energy life cycle design: A method. *J. Eng. Manuf.* 1999, *213*, 415–419.

50. Despeisse, M.; Mbaye, F.; Ball, P.D.; Evans, S.; Levers, A. Emergence of sustainable manufacturing practices. *Int. J. Prod. Plan. Control* 2012, *23*, 354–376.

Chapter 10

ECOLOGY AND ESCIENCE

Christian Mulder
National Institute for Public Health and the Environment (RIVM)

COMPREHENDING THE GIFTS OF ECOLOGY

Stress ecology, climate change, human well-being, and global sustainability are popular items (Naeem et al. 2009). Given all the challenges in a developing world where the global population is supposed to reach 9.15 billion in 2050 (Pimentel et al. 1999; United Nations Population Division 2010), policy makers are, for the first time, keen on concrete assessments of our world, looking with interest and fear to ecological models. Although the discussion between scientists and politicians is known to be difficult, too many recent catastrophes during a single year - from the British Petroleum oil spill in the Gulf of Mexico (De Gouw et al. 2011) up to the ongoing radioactive Fukushima wreckage (Schiermeier 2011) - rapidly forced a better and constructive interaction between applied ecologists and policy makers at different organization levels. Such an interaction is also reflected by the arising use of internet metrics, blogs, tweets, and social networking - all digital tools that are already more or less linked to the thought process that society and policy are currently going through. Scientists are used to the Web of Science for selecting the appropriate papers, and policy makers are using methodologies for weighing opinions (Bollen et al. 2009). The latter authors even defined modern science as a 'gift economy', and they are absolutely right. What else should happen to improve the interactions between policy and research?

eScience: Computation Impacted Science

As can be seen in the plethora of data and the huge degree of multidisciplinarity (Figure 1), *a full and immediate open access across all disciplines* is supposed to be the dream of any eScientist and stakeholder involved in this thought process. At a glance, ecologists should be eScientists *par excellence*. Most ecological disciplines, in fact, overlap and typically (re)use data from other sciences, which leads to a huge increase in science productivity. Ecology often benefits from methods originally developed for mathematics, physics, and chemistry (Cohen 2004; Elser 2006) and might even benefit from models used in the worlds of informatics and finance (Allesina and Pascual 2009; Haldane and May 2011, respectively). Opposite flows occur as well, from ecological knowledge to archaeology and urban systems, toxicology, and medicine (Sjögren and Lamentowicz 2008; Grimm et al. 2000; Van Straalen 2003; Zu Dohna et al. 2009, respectively). Ecology is a perfect eScience, and I believe that no scientific discipline has already so many examples of mutual benefits as all ecological (sub)disciplines: molecular/evolutionary ecology and genetics, ecology and economics, ecology and geology, ecology and paleontology, ecology and atmospheric deposition, ecology and climate regulation, ecology and environmental planning, and ecology and environmental quality (e.g.,Dicke et al. 2011; Costanza et al. 1997; Dommain et al. 2011; Dunne et al. 2008; Elser et al. 2009; West et al. 2011; Taylor Lovell and Johnston 2009; Mulder et al. 2011, respectively). Although most attention is provided to pristine terrestrial biota and extreme environments (e.g., Bai et al. 2004; Yu et al. 2010), ecology provides the necessary knowledge to assess consequences of disturbance and stress, such as the effects of pollution, mining, and engineering on local communities at the landscape level (De Zwart et al. 2006; Dugan et al. 2010; Wang et al. 2010). Moreover, using several case studies from the Great Barrier Reef, Stoeckl et al. (2011) pointed out how important it is to differentiate between information about existing ecosystem services (Adhikari and Nadella 2011) and information about the extent to which existing ecosystem services might change (or have already changed) in response to external events (West et al. 2011).

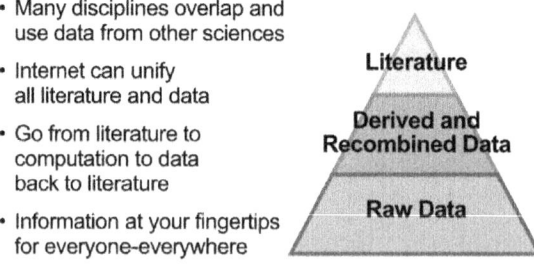

Figure 1: eScience is about global collaboration: a world where data and literature interoperate with each other. (Adapted from Hey and Trefethen 2003; Henzinger and

Lawrence 2004, and references therein; photo credit by Hey et al. 2009; free eBook at http://fourthparadigm.org, courtesy of Microsoft Research.).

The World of Science Changed Radically

Science is at a tipping point at which research will rapidly become more applied. Land conversion for agriculture is projected to rise sharply (Laurance 2001), and strong empirical evidence is available. Thanks to the huge increase in peer-reviewed publications, ecology helps in understanding what is going on, in sustaining current ecosystems and in managing novel ecosystems, and contributes in the societal thought process. Indeed, nobody is questioning that the entire world of science has changed (Laurance 2001; Hey et al. 2009). From this perspective, open-access journals (ranging from cosmology to environment and from medicine to biology) are for sure the best way to share empirical data, models, and thoughts with others (Bourne et al. 2008; Figure 2). Successful linking of (e)Science with decision making will depend on accuracy and communication (Clark et al. 2001).

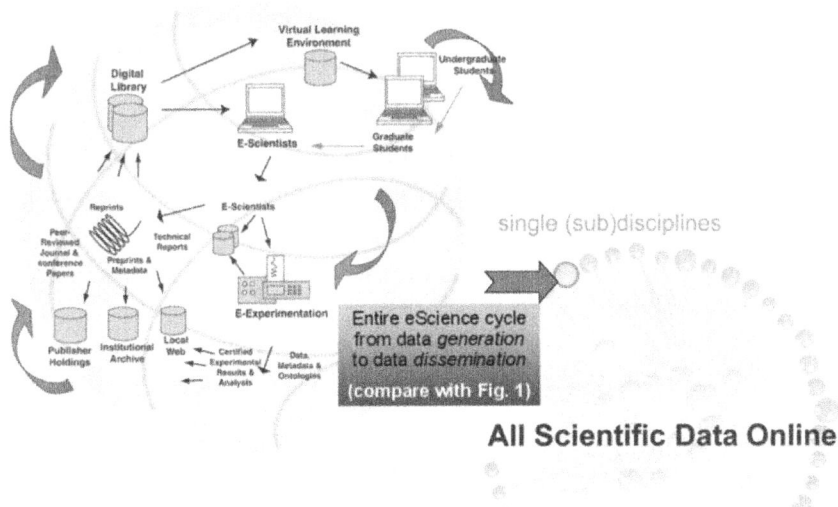

Figure 2: The entire eScience cycle from data generation to data dissemination. eScience remains linked to the requirements for dataBank procedures, i.e., the quality, integrity, identity, provenance, description, and ontology of original scientific data. (Grid photo courtesy of Liz Lyon, e-Bank, UK; modified from Hey and Trefethen 2005; network credit by Alice Boit, University of Potsdam.).

Balancing theory and empirical data at different levels allowed a shift in the scientific inquiries, resulting in an improved performance of current investigation. (Previously, a myriad of topics made navigation more difficult.) Internet, networking, international programs, and the astonishingly high amount of data confluence into this new conceptual framework (Hey et al. 2009), where any traditional (sub)discipline has been forced to change (see Box 1 and references therein; Mulder et al. 2011). Internet can unify such (sub) disciplines in a fractal-like world with increasing complexity and accessible detail (Figure 2), demonstrating the need to unify philosophy of science with the science and technology studies.

Albeit some ecologists seem not to be fully aware of the enhancements of online scholarly communication, open access is really beginning to spread. Still, only one-fourth of the contributions published between 2003 and 2010 by the Public Library of Science [PLoS] (2011) belong to the categories Development and Evolutionary Biology, Ecology, Marine and Aquatic Sciences, or Plant Biology, and less than 0.15% of the seven hundreds of thousand contributions deposited during the same period in the arXiv (2011) belong to the 'Quantitative Biology' category. Seeing how many ecologists appreciate PLoS (and open access in general), these relatively low contributions surprise me. Something comparable occurs with data sharing: empirical data sets are of the highest value (*cf.* Bourne 2005), but ecologists seem to become 'shy' when asked to make their data accessible. This makes me question why in 2010, only 7% of all the participants to one EU-funded project responded to their open data call?

This short overview shows, on the one hand, how many ecologists are already dealing with data sharing and eScience, and on the other hand, how different complementary frameworks seem to coexist. The overlap between schools and frameworks can be rather confounding. For instance, although many features are common to different ways of performing research, metadata focusing on compiled information for local communities are widespread as ecological networks, whereas scientific data derived from simulated or empirical information are commonly defined as food webs. This implies that webs can, in most cases, be seen as networks although not all ecological networks may be seen as traditional food webs. However, in contrast to many nested networks, eScience is mutual and, as such, is expected to grow with vigor because, in contrast to (antagonistic) living organisms, eScience has no discrete boundaries but has digitally cross-connected domains with high modularity.

Thinking Seriously About the Scenario

After all the efforts of producing a research paper (from laboratory and/or field work up to forecasting, networking, feedbacks, and final writing), how are ecologists currently thinking about their outreach? Bourne (2010) provocatively wrote that in contrast to the rather static PDF interface, publishing workflows - and preprint repositories and post-publication commentaries can be seen as a kind of creative workflows - are more powerful but harder to manage and represent, therefore, *a major change for most scientists*.

Apparently, ecologists belong to those scientists, but if they are not able to keep online the attention of their own colleagues, then how can they expect any concrete feedbacks from environmentally interested laypersons and policy decision makers? How can stakeholders expect a concrete use of ecological indicators if most scientists dealing with such an integrative discipline seem to (be willing to) ignore a large part of data and recent literature on environmental impact and ecosystem services (*cf.* Bjorndal et al. 2011)?

Having seen these disputes between institutions, it is not a wonder that Al Gore has succeeded in areas where so many scientists, stakeholders, and NGOs have failed. It is time for a radical departure; to improve the synergy between authors, publishers, and readers, two possible incentives should be considered as novel opportunities:

- Be explicit and do not try to be exhaustive

There is an increasing amount of review papers cited to support empirical data, and most results can be interesting for a wide variety of scientists if the original results could be put in a much wider context. Most manuscripts just aim to bridge knowledge gaps, but overwhelming references are always inappropriate to catch real attention, and chains of citations have to be avoided as they move from journal to journal (Hirsch 2005). We should strive to generate less academic papers and attempt to sample the web uniformly.

- Be wide and do not claim something else

Bibliographic couplings aim to identify structures and communities (Menczer 2004). To avoid strong discipline bias, take substantial precaution. Do not claim shocking lack of knowledge (such terms are not objective) and do not cite only the papers that seem to confirm your results (Jennions and Møller 2002). Your literature shows your horizons. In the framework of eScience, papers always benefit from testing the theory or hypothesis regarding how relationships change, so be wide in the chosen references.

Let us put matters straight: ecological forecasts are imperative (Clark et al. 2001). These authors defined ecological forecasting as 'the process of predicting the state of ecosystems, ecosystem services, and natural capital' (Clark et al. 2001). Although some scientists still prefer to claim that no data are available to specify uncertainties or find that novel ideas cannot be tested properly for lack of appropriate technology (*cf.* Collins 2010), others state that larger data sets (inclusive historic data sets) are meanwhile discoverable (Hunt et al. 2009), making ecological forecasting feasible. Data are interwoven with peer-reviewed scientific papers, are authored in digital form, and are benefits of today's structural use of the entire digital environment (Lynch 2009).

A New Scientific Endeavor Online

As a matter a fact, 'science aims to produce far more than a simple mechanical prediction of correlations' (Ginsparg 2009), and most open-access journals clearly aim to support a better *comprehension* of the exabytes of already available information and to improve strongly the*dissemination* of data, causalities, and implications. Therefore, *Ecological Processes* is governed by three principles: research quality, multidisciplinarity and integration, and open access. (Overarching principles are reflected by all those who have agreed to provide part of their precious time as committed editors.) As editors in chief, we all have plenty of ideas for what *Ecological Processes* means, but we really wish to get your input too, as you like it.

REFERENCES

1. Adhikari B, Nadella K: Ecological economics of soil erosion: a review of the current state of knowledge. *Ann New York Acad Sci*2011, 1219: 134–152. 10.1111/j.1749-6632.2010.05910.x
2. Allesina S, Pascual M: Googling food webs: can an Eigenvector measure species› importance for coextinctions? *PLoS Comput Biol*2009,5(9):e1000494. 10.1371/journal.pcbi.1000494
3. arXiv Cornell University Library, Ithaca, NY; 2011.http://arxiv.org . Accessed 20 Apr 2011
4. Bai Y, Han X, Wu J, Chen Z, Li L: Ecosystem stability and compensatory effects in the Inner Mongolia grassland. *Nature* 2004, 431:181–184. 10.1038/nature02850
5. Bjorndal KA, Bowen BW, Chaloupka M, Crowder LB, Heppell SS, Jones CM, Lutcavage ME, Policansky D, Solow AR, Witherington BE:Better science needed for restoration in the Gulf of Mexico. *Science* 2011, 331: 537–538. 10.1126/science.1199935
6. Bollen J, Van de Sompel H, Hagberg A, Chute R: A principal component analysis of 39 scientific impact measures. *PLoS ONE*2009,4(6):e6022. 10.1371/journal.pone.0006022
7. Bourne P: Will a biological database be different from a biological journal? *PLoS Comput Biol* 2005,1(3):e34. 10.1371/journal.pcbi.0010034

8. Bourne PE: What do I want from the publisher of the future? *PLoS Comput Biol* 2010,6(5):e1000787. 10.1371/journal.pcbi.1000787
9. Bourne PE, Fink JL, Gerstein M: Open access: taking full advantage of the content. *PLoS Comput Biol* 2008,4(3):e1000037. 10.1371/journal.pcbi.1000037
10. Clark JS, Carpenter SR, Barber M, Collins S, Dobson A, Foley JA, Lodge DM, Pascual M, Pielke R Jr, Pizer W, Pringle C, Reid WV, Rose KA, Sala O, Schlesinger WH, Wall DH, Wear D: Ecological forecasts: an emerging imperative. *Science* 2001, 293: 657–660. 10.1126/science.293.5530.657
11. Cohen JE: Mathematics is biology›s next microscope, only better; biology is mathematics› next physics, only better. *PLoS Biol* 2004,2(4):e439.
12. Collins JP: Sailing on an ocean of 0s and 1s. *Science* 2010, 327: 1455–1456. 10.1126/science.1186123
13. Costanza R, D›Arge R, De Groot R, Farber S, Grasso M, Hannon B, Limburg K, Naeem S, O›Neill RV, Paruelo J, Raskin RG, Sutton P, Van den Belt M: The value of the world›s ecosystem services and natural capital. *Nature* 1997, 387: 253–260. 10.1038/387253a0
14. De Gouw JA, Middlebrook AM, Warneke C, Ahmadov R, Atlas EL, Bahreini R, Blake DR, Brock CA, Brioude J, Fahey DW, Fehsenfeld FC, Holloway JS, Le Henaff M, Lueb RA, McKeen SA, Meagher JF, Murphy DM, Paris C, Parrish DD, Perring AE, Pollack IB, Ravishankara AR, Robinson AL, Ryerson TB, Schwarz JP, Spackman JR, Srinivasan A, Watts LA: Organic aerosol formation downwind from the Deepwater Horizon oil spill. *Science* 2011, 331: 1295–1299. 10.1126/science.1200320
15. De Zwart D, Dyer SD, Posthuma L, Hawkins CP: Predictive models attribute effects on fish assemblages to toxicity and habitat alteration. *Ecol Appl* 2006, 16: 1295–1310. 10.1890/1051-0761(2006)016[1295:PMAEOF]2.0.CO;2
16. Dicke M, Van Loon JJA, De Jong PW: Ecogenomics benefits community ecology. *Science* 2011, 305: 618–619.
17. Dommain R, Couwenberg J, Joosten H: Development and carbon sequestration of tropical peat domes in south-east Asia: links to post-glacial sea-level changes and Holocene climate variability. *Quater Sci Rev* 2011, 30: 999–1010. 10.1016/j.quascirev.2011.01.018
18. Dugan PJ, Barlow C, Agostinho AA, Baran E, Cada GF, Chen D, Cowx IG, Ferguson JW, Jutagate T, Mallen-Cooper M, Marmulla G, Nestler J, Petrere M, Welcomme RL, Winemiller KO: Fish migration, dams, and loss of ecosystem services in the Mekong basin.*Ambio* 2010, 39: 344–348. 10.1007/s13280-010-0036-1

19. Dunne JA, Williams RJ, Martinez ND, Wood RA, Erwin DH: Compilation and network analyses of Cambrian food webs. *PLoS Biol* 2008,6(4):e102. 10.1371/journal.pbio.0060102

20. Elser JJ: Biological stoichiometry: a chemical bridge between ecosystem ecology and evolutionary biology. *Am Nat* 2006, 168:S25-S35. 10.1086/509048

21. Elser JJ, Andersen T, Baron JS, Bergström A-K, Jansson M, Kyle M, Nydick KR, Steger L, Hessen DO: Shifts in lake N:P stoichiometry and nutrient limitation driven by atmospheric nitrogen deposition. *Science* 2009, 326: 835–837. 10.1126/science.1176199

22. Ginsparg P: Scholarly communication. In *The fourth paradigm: data-intensive scientific discovery. Redmond, Microsoft Research* Edited by: Hey T, Tansley S, Tolle K. 2009, 185–191.

23. Grimm NB, Grove JM, Pickett STA, Redman CL: Integrated approaches to long-term studies of urban ecological systems.*BioScience* 2000, 50: 571–584. 10.1641/0006-3568(2000)050[0571:IATLTO]2.0.CO;2

24. Haldane AG, May RM: Systemic risk in banking ecosystems. *Nature* 2011, 469: 351–355. 10.1038/nature09659

25. Henzinger M, Lawrence S: Extracting knowledge from the World Wide Web. *Proc Natl Acad Sci USA* 2004, 101: 5186–5191. 10.1073/pnas.0307528100

26. Hey T, Trefethen A: e-Science and its implications. *Phil Trans R Soc Lond A* 2003, 361: 1809–1825. 10.1098/rsta.2003.1224

27. Hey T, Trefethen A: Cyberinfrastructure for e-Science. *Science* 2005, 308: 817–821. 10.1126/science.1110410

28. Hey T, Tansley S, Tolle K (Eds): The fourth paradigm: data-intensive scientific discovery In *Microsoft Research, Redmond* 2009.

29. Hirsch JE: An index to quantify an individual›s scientific research output. *Proc Natl Acad Sci USA* 2005, 102: 16569–16572. 10.1073/pnas.0507655102

30. Hunt JR, Baldocchi DD, Van Ingen C: Redefining ecological science using data. In *The fourth paradigm: data-intensive scientific discovery. Microsoft Research, Redmond* Edited by: Hey T, Tansley S, Tolle K. 2009, 21–26.

31. Jennions MD, Møller AP: Publication bias in ecology and evolution: an empirical assessment using the ‹trim and fill› method. *Biol Rev* 2002, 77: 211–222.

32. Laurance WF: Future shock: forecasting a grim fate for the Earth. *Trends*

Ecol Evol 2001, 16: 531–533. 10.1016/S0169-5347(01)02268-6

33. Lynch C: Jim Gray›s fourth paradigm and the construction of the scientific record. In *The fourth paradigm: data-intensive scientific discovery. Microsoft Research, Redmond* Edited by: Hey T, Tansley S, Tolle K. 2009, 177–183.

34. Menczer F: Evolution of document networks. *Proc Natl Acad Sci USA* 2004, 101: 5261–5265. 10.1073/pnas.0307554100

35. Mulder C, Boit A, Bonkowski M, De Ruiter PC, Mancinelli G, Van der Heijden MGA, Van Wijnen HJ, Vonk JA, Rutgers M: A belowground perspective on Dutch agroecosystems: how soil organisms interact to support ecosystem services. *Adv Ecol Res* 2011, 44: 277–357.

36. Naeem S, Bunker DE, Hector A, Loreau M, Perring CP (Eds): *Biodiversity, ecosystem functioning, and human wellbeing* Oxford University Press, Oxford; 2009.

37. Pimentel D, Bailey O, Kim P, Mullaney E, Calabrese J, Walman L, Nelson F, Yao X: Will limits of the Earth›s resources control human numbers? *Environ Dev Sustain* 1999, 1: 19–39. 10.1023/A:1010008112119

38. PLoS Public Library of Science, San Francisco, CA; 2011.http://www.plos.org . Accessed 20 Apr 2011

39. Schiermeier Q: Radiation release will hit marine life. *Nature* 2011, 472: 145–146. 10.1038/472145a

40. Sjögren P, Lamentowicz M: Human and climatic impact on mires: a case study of Les Amburnex mire, Swiss Jura Mountains. *Veg Hist Archaeobot* 2008, 17: 185–197. 10.1007/s00334-007-0095-9

41. Stoeckl N, Hicks CC, Mills M, Fabricius K, Esparon M, Kroon F, Kaur K, Costanza R: The economic value of ecosystem services in the Great Barrier Reef: our state of knowledge. *Ann New York Acad Sci* 2011, 1219: 113–133. 10.1111/j.1749-6632.2010.05892.x

42. Taylor Lovell S, Johnston DM: Creating multifunctional landscapes: how can the field of ecology inform the design of the landscape? *Front Ecol Environ* 2009, 7: 212–220. 10.1890/070178

43. United Nations Population Division: World population prospects. 2010. http://esa.un.org/unpd/wpp/index.htm. Accessed 18 Jan 2012

44. Van Straalen NM: Ecotoxicology becomes stress ecology. *Environ Sci Technol* 2003, 37: 324A-330A. 10.1021/es0325720

45. Wang J, Huang J, Wu J, Han X, Lin G: Ecological consequences of the Three Gorges Dam: insularization affects foraging behavior and dynamics of rodent populations. *Front Ecol Environ* 2010, 8: 13–19.

10.1890/070188

46. West PC, Narisma GT, Barford CC, Kucharik CJ, Foley JA: An alternative approach for quantifying climate regulation by ecosystems. *Front Ecol Environ* 2011, 9: 126–133. 10.1890/090015

47. Yu Q, Chen Q, Elser JJ, He N, Wu H, Zhang G, Wu J, Bai Y, Han X: Linking stoichiometric homeostasis with ecosystem structure, functioning, and stability. *Ecol Lett* 2010, 13: 1390–1399. 10.1111/j.1461-0248.2010.01532.x

48. Zu Dohna H, Cecere MC, Gürtler RE, Kitron U, Cohen JE: Spatial re-establishment dynamics of local populations of vectors of Chagas disease. *PLoS Negl Trop Dis* 2009,3(7):e490. 10.1371/journal.pntd.0000490

Chapter 11

A BIBLIOMETRIC ANALYSIS OF GLOBAL FOREST ECOLOGY RESEARCH DURING 2002-2011

Yajun Song[1,2] and Tianzhong Zhao[1]

[1]School of Information Science & Technology, Beijing Forestry University

[2]Library of Beijing International Studies University

ABSTRACT

Bibliometric is increasingly used for the analysis of discipline dynamics and management related decision-making. This study analyzes 937,923 keywords from 78,986 articles concerning forest ecology and conducts a serial analysis of these articles' characteristics. The articles' records, published between 2002 and 2011, were downloaded from the Web of Science, and their keywords were exported by Java processing programs. The result shows that forest ecology studies focused on forest diversity, conservation, dynamics and vegetation in the last decade. Developed countries, such as the USA, Canada, and Germany, were the most productive countries in the field of forest ecology research. From 2002 to 2011, the number of articles published annually related to forest ecology grew at a stable rate, as indicated by the fit produced by a high determination coefficient ($R^2 = 0.9955$). The findings of this study may be applicable for planning and managing forest ecology research and partners involved in such research may use this study as a reference.

INTRODUCTION

Bibliometric analysis is an important part of reference and research services. Forest ecology is closely related to forest management and many studies have been performed from various perspectives, including studies of ecosystems at multiple forest spatial scales (Rodrigues et al. 2011;Sitzia et al. 2010), long term ecosystem change (Diaz et al. 2007;van Oudenhoven et al. 2012), climate change (Cheaib et al. 2012; Şekercioğlu et al.2012), soils (McLachlan and Bazely2003; Wang et al.2011), physiography (Morrissey et al.2009; Rubio and Escudero2005), carbon balance (Mitchell et al.2009; Sillett et al.2010), nutrient cycling (Berger et al.2009; XU and Chen2006), landscape ecology (Loucks et al.2001; Wintle et al.2005) and biodiversity (Hanberry et al.2012; Lamb et al.2005). In addition to these studies, a bibliometric analysis of global forest ecology could provide a fresh look at the current status of global forest ecology research and help identify hot spots.

In recent years, along with its continuously expanding range of application, bibliometric analysis plays an increasingly important role in management and decision-making in science and technology. It has been used to document the development of some research fields (Grandjean et al.2011; Hendrix2008; Narotsky et al.2012; van Eck et al.2010; van Raan2006), including forestry (Dobbertin and Nobis2010; Perez et al.2004).

In this study, we perform a bibliometric analysis of forest ecology research over the last 10 years (2002–2011) aimed at (1) examining the temporal hot topics of forest ecology research by keyword frequency analysis, (2) revealing the distribution of articles by country/region, organization, funding agency, research area, author, year and publication name for articles covering forest ecology research and revealing advancements in forest ecological research, and (3) providing a new keywords frequency analysis method, which may benefit future research.

MATERIALS AND METHODOLOGY

Data Collection

Literature records, our analytical objects, were derived from the Web of Science, an online academic citation index database provided by Thomson Reuters. To define search terms, we used the "thesaurus" tool of Commonwealth Agricultural Bureaux (CAB) Abstracts.

We conducted a search on the word "ecology" in CAB Abstracts and the search produced 41 terms, including 19 narrower terms and 22 other related terms (Figure 1). We selected terms with more than 200 hits and used Microsoft Excel to rank them in descending order. We then removed the words "ecology" and "forest" from the Excel sheet and added the terms "climate," "soils," "physiography," "carbon balance" and "nutrient cycling," based on the concepts related to forest ecology defined by Barnes et al. (1997). Then, we defined the remaining 43 search terms and constructed a new search query. The search was limited to "article" type publications published between 1 January 2002 and 31 December 2011 in English.

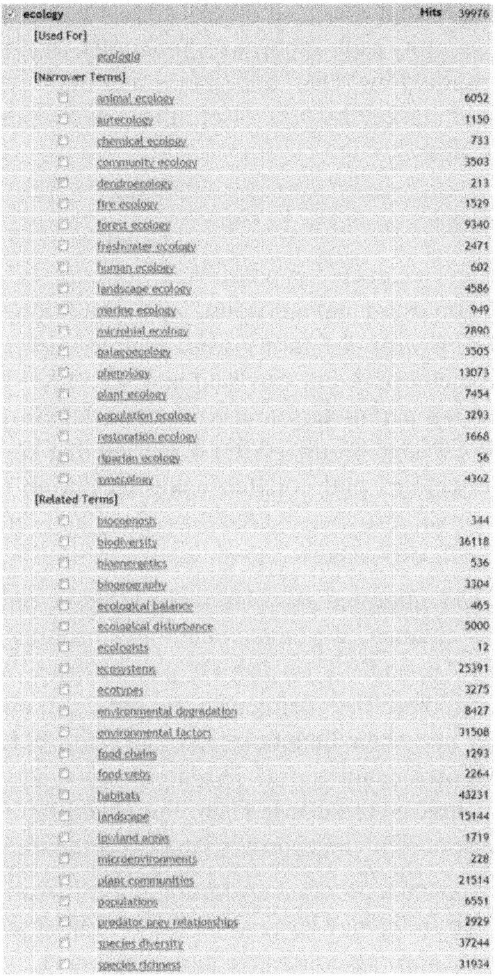

ecology	Hits 39976
[Used For]	
ecologia	
[Narrower Terms]	
animal ecology	6052
autecology	1150
chemical ecology	733
community ecology	3503
dendroecology	213
fire ecology	1529
forest ecology	9340
freshwater ecology	2471
human ecology	602
landscape ecology	4586
marine ecology	949
microbial ecology	2890
palaeoecology	3305
phenology	13073
plant ecology	7454
population ecology	3293
restoration ecology	1668
riparian ecology	56
synecology	4362
[Related Terms]	
biocoenosis	344
biodiversity	36118
bioenergetics	536
biogeography	3304
ecological balance	465
ecological disturbance	5000
ecologists	12
ecosystems	25391
ecotypes	3275
environmental degradation	8427
environmental factors	31508
food chains	1293
food webs	2264
habitats	43231
landscape	15144
lowland areas	1719
microenvironments	228
plant communities	21514
populations	6551
predator prey relationships	2929
species diversity	37244
species richness	31934

Figure 1: Narrower terms and 22 related terms of ecology.

The search query included 43 terms (see Appendix A). This query was run in Web of Science, which is a citation database of the Web of Knowledge, and a total of 78,986 forest ecology-related articles were identified.

Using the Web of Science's analysis tools, we exported the 78,986 articles by country/region, organization, funding agency, research area, author, year, and publication. The statistical methods used by the Web of Science for the above statistical indicators of multi-author articles do not distinguish between the order of author's locations, which may result the sum of these statistical result was greater than 78,986. The article records, including title, author, keywords, abstract, and organization, were exported in full record mode from the Web of Science to text files. A total of 158 text files were created, because the Web of Science limits each export to 500 records. In every text file, "author keywords" were marked by "DE," and "keywords plus" were provided by the Web of Science and marked by "ID". Both these two kinds of keywords were considered in this study.

Keywords Analysis

First, the frequency of each keyword was counted in each text file. We developed a java program named count.java (Additional file 1: Appendix B) using Eclipse software, a famous cross-platform integrated development environment. This java program can find and select keywords in the output text file by identifying parameters, and connect each keyword to a long string, while deleting the carriage returns. After detection, the keywords in the string were split by semicolons, and counted using HashMap traversal algorithm. The HashMap traversal result was saved to an array and sorted by the counters; then, the sorted result was exported to an intermediate file.

Second, the 158 intermediate files were merged, and the frequency of each keyword was counted. We developed a java program named merge.java (Additional file 1: Appendix C) using Eclipse software. When this program was run, the intermediate files defined in the input parameters were opened, and the keywords and their counters were saved to a HashMap. Then the keywords were counted again with HashMap traversal algorithm: the counters of the same keywords were added. Then, the HashMap traversal result was saved to an array, sorted by the counters, and exported into a result file.

Third, we developed a program (Additional file 1: Appendix D) to create a java package named frequency.jar to store the compiled java class files which were produced by compiling count.java and merge.java.

Fourth, we developed a batch program named count.bat (Additional file 1: Appendix E) to call the count.class with the input parameters "DE" and "ID". All 158 text files were processed one by one. As a result, 158 intermediate files were created.

Fifth, we developed another batch program named merge.bat (Additional file 1: Appendix F) to call the merge.class with the input parameters, that is, the 158 intermediate files, to merge them. As a result, a final file was created, in which all keywords in 78,986 articles were counted and sorted.

After data processing, 937,923 keywords from those 78,986 articles were merged into 150,974 keywords. All of the keywords were sorted in reverse order based on their frequencies. The 100 most frequently used keywords became the focus of our study.

RESULTS

Keywords Analysis Results

To narrow the research scope, the 100, 200, 300 most frequently used keywords were selected and analyzed. As a result, the 100 most frequently used keywords, 0.07% of the 150,974 unique keywords analyzed here represented 18.54% of the total (937,923) of all keywords harvested (Table 1). We focused on the top 100 keywords to examine the hot topics of forest ecology research (Table 2).

Table 1: The top 100, 200, 300 keyword ratio and their frequencies

Keywords number	Keywords ratio	Keywords frequencies	Frequencies Ratio (%)
100	0.07%(100/150974)	173925	18.54%(173925/937923)
200	0.13%(200/150974)	233042	24.85%(233042/937923)
300	0.20%(300/150974)	271233	28.92%(271233/937923)

Table 2: The top 100 keywords in forest ecology articles indexed using the Web of Science during 2002–2011

	Keywords	Frequencies
1	forest	9302
2	diversity	5424
3	conservation	5135
4	dynamics	4886
5	vegetation	4720
6	biodiversity	4613
7	patterns	4166
8	growth	4069
9	rain-forest	3253
10	management	3236
11	nitrogen	3136
12	forests	3069
13	soil	2793
14	ecology	2677
15	communities	2596
16	carbon	2568
17	climate-change	2412

18	ecosystems	2407
19	disturbance	2389
20	species richness	2381
21	boreal forest	2334
22	landscape	2180
23	biomass	2130
24	model	2100
25	climate	2095
26	fire	2043
27	abundance	1855
28	united-states	1849
29	habitat	1846
30	temperature	1824
31	plants	1782
32	organic-matter	1755
33	populations	1733
34	decomposition	1603
35	climate change	1599
36	dispersal	1590
37	responses	1576
38	regeneration	1531
39	tropical forest	1513
40	land-use	1509
41	habitat fragmentation	1495
42	trees	1486
43	fragmentation	1473
44	forest soils	1441
45	evolution	1408
46	succession	1384
47	deforestation	1375
48	ecosystem	1362
49	birds	1333
50	population	1276
51	competition	1273
52	water	1235

53	variability	1210
54	deciduous forest	1190
55	forest management	1189
56	community structure	1178
57	behavior	1140
58	community	1131
59	restoration	1127
60	tropical forests	1107
61	photosynthesis	1093
62	seed dispersal	1081
63	usa	1067
64	productivity	1054
65	microbial biomass	1040
66	density	1034
67	impact	1019
68	brazil	1018
69	models	988
70	carbon-dioxide	978
71	phosphorus	971
72	size	971
73	predation	947
74	classification	943
75	respiration	932
76	scale	927
77	drought	920
78	national-park	918
79	plant	910
80	selection	909
81	tree	902
82	deposition	889
83	history	888
84	recruitment	875
85	norway spruce	874

86	soil respiration	870
87	australia	868
88	consequences	864
89	tropical rain-forest	839
90	survival	834
91	quality	830
92	mexico	819
93	costa-rica	813
94	impacts	812
95	new-zealand	796
96	forest soil	794
97	mortality	788
98	soils	787
99	grassland	786
100	assemblages	785

Articles Analysis Result

By Country/Region

The 78,986 articles were analyzed by countries or regions and sorted in reverse order by their total numbers and Table 3 lists the results for the top 20 countries. We supplemented a column in the original table and classified these 20 countries/regions by their respective continents, which showed that North America and 12 European countries had about 44.71% and 42.35% of all the articles, respectively, indicating published articles related to forest ecology in North America and Europe predominate.

Table 3: Top 20 countries/regions publishing articles on forest ecology indexed using the web of science during 2002–2011

	Countries/Regions	Records	Ratio (%)	Continents
1	USA	28060	35.53	North America
2	Canada	7255	9.19	North America
3	Germany	6311	7.99	Europe
4	Brazil	4561	5.77	Africa
5	Australia	4375	5.54	Australia

6	England	4229	5.35	Europe
7	Peoples R China	4122	5.22	Asia
8	France	3930	4.98	Europe
9	Japan	3504	4.44	Asia
10	Spain	3402	4.31	Europe
11	Sweden	2708	3.43	Europe
12	Finland	2417	3.06	Europe
13	Italy	2230	2.82	Europe
14	Netherlands	1921	2.43	Europe
15	Switzerland	1871	2.37	Europe
16	India	1798	2.28	Asia
17	Mexico	1572	1.99	South America
18	Russia	1554	1.97	Europe
19	Scotland	1455	1.84	Europe
20	New Zealand	1421	1.80	Europe

The combined frequency of keywords related to tropical forest, represented by "rain-forest" (3,253), "tropical forest" (1,513), "tropical forests" (1,107), and "tropical rain-forest" (839), totaled 6,712 keyword entries, which was exceeded only by the keyword "forest" with 9,302 entries (Table 2). This indicates that tropical forest was the main focus of research in forest ecology studies. Tropical forest is mainly distributed in Southeast Asia, Central America, South America, Australia, Africa. However, the main countries with strong research capabilities related to tropical forest research were not located in those areas, but were found in North America and Europe.

By Organization

Forest ecology studies were conducted by 7,598 organizations, and Table 4 lists the top 20 organizations and their related countries. The University of California System, the Chinese Academy of science, and US Forest Service produced the most articles. Eight organizations were from the USA, two each from Canada, Brazil, and Germany, and the remaining six were from China, Sweden, Finland, Russia, Spain, and France.

Table 4: Top 20 organizations publishing articles on forest ecology indexed using the web of science during 2002–2011

	Organizations	Records	Ratio (%)	Counties
1	Univ Calif System	2749	3.48	USA

2	Chinese Acad SCI	2359	2.99	China
3	US Forest Serv	2203	2.79	USA
4	Swedish Univ Agr SCI	1342	1.70	Sweden
5	Oregon State Univ	1200	1.52	USA
6	Univ Helsinki	1055	1.34	Finland
7	Univ British Columbia	1008	1.28	Canada
8	Univ Wisconsin System	978	1.24	USA
9	Univ Alberta	973	1.23	Canada
10	Russian Acad SCI	925	1.17	Russia
11	Univ Florida	905	1.15	USA
12	USDA	905	1.15	USA
13	Univ Sao Paulo	896	1.13	Brazil
14	US Geol Survey	883	1.12	USA
15	Univ Fed Santa Maria	868	1.10	Brazil
16	Smithsonian Inst	867	1.10	USA
17	Max Planck Society	808	1.02	Germany
18	Univ Gottingen	785	0.99	Germany
19	INRA	771	0.98	France
20	CSIC	766	0.97	Spain

USDA United States department of agriculture, *INRA* Institut National de la recherche agronomique, *CSIC* consejo superior de investigaciones científicas.

By Funding Agency

6,356 funding agencies subsidized forest ecology studies, and the top 20 were exported for closer analysis. Because many articles used abbreviations for the funding agencies the top 20 were merged into 15 (Table 5). Examples include the National Science Foundation (NSF), the Conselho Nacional de Desenvolvimento Científico e Tecnológico (CNPq), the European Union (EU), and the Natural Sciences and Engineering Research Council of Canada (NSERC).

Table 5: The 15 most productive agencies funding forest ecology research indexed by the web of science during 2002–2011

	Funding agencies	Articles number	Ratio (%)	Countries
1	National Science Foundation	2240	2.84	USA

2	National Natural Science Foundation of China	831	1.05	China
3	Natural Sciences and Engineering Research Council of Canada	807	1.02	Canada
4	CNPq	744	0.94	Brazil
5	European Union	601	0.76	EU
6	Chinese Academy of Sciences	372	0.47	China
7	NASA	357	0.45	USA
8	European Commission	337	0.43	EC
9	Academy of Finland	311	0.39	Finland
10	Australian Research Council	265	0.34	Australia
11	CAPES	221	0.28	Brazil
12	National Basic Research Program of China	196	0.25	China
13	FAPESP	192	0.24	Brazil
14	Russian Foundation for Basic Research	185	0.23	Russia
15	USDA Forest Service	172	0.22	USA

CNPQ: Conselho Nacional de Desenvolvimento Científico e Tecnológico, Brazil; NASA: National Aeronautics and Space Administration, USA; CAPES: Coordenação de Aperfeiçoamento de Pessoal de Nivel Superior, Brazil; FAPESP: Fundação de Amparo à Pesquisa do Estado de São Paulo, Brazil; EC: European Commission.

The National Science Foundation (USA), National Natural Science Foundation of China (China), Natural Sciences and Engineering Research Council of Canada (Canada), Conselho Nacional de Desenvolvimento Científico e Tecnológico (Brazil), and European Commission were more prolific in forest ecology than other funding agencies. Combining the number of articles in Table 5 by country/region demonstrates that the USA (2,769), China (1,399), Brazil (1,157), Canada (807), and EU (601) were also the top five countries/regions and provided more financial aid to forest ecology research than other countries.

By Research Area

In the analysis, forest ecology was related to 72 research areas identified by the Web of Science data. Table 6 lists the top 20 research areas and clearly shows that forest ecology studies were related to a wide range of disciplines. Environmental sciences ecology (31,172 or 39.47% of all articles), forestry

(13,164, 16.67%), agriculture (8,354, 10.58%), and plant sciences (8,027, 10.16%) were the top four major related research areas.

Table 6: The top 20 research areas related to forest ecology indexed using the web of science during 2002–2011

	Research areas	Articles number	Ratio (%)
1	Environmental Sciences Ecology	31172	39.47
2	Forestry	13164	16.67
3	Agriculture	8354	10.58
4	Plant Sciences	8027	10.16
5	Zoology	6470	8.19
6	Biodiversity Conservation	6005	7.60
7	Geology	5660	7.17
8	Meteorology Atmospheric Sciences	3654	4.63
9	Physical Geography	3453	4.37
10	Water Resources	2521	3.19
11	Marine Freshwater Biology	2271	2.88
12	Entomology	2176	2.76
13	Engineering	1981	2.51
14	Life Sciences Biomedicine Other Topics	1650	2.09
15	Evolutionary Biology	1631	2.07
16	Remote Sensing	1611	2.04
17	Science Technology Other Topics	1319	1.67
18	Biochemistry Molecular Biology	1269	1.61
19	Imaging Science Photographic Technology	1205	1.53
20	Genetics Heredity	1079	1.37

By Author

A total of 48,373 authors participated in forest ecology related studies. Among the 20 authors publishing the most articles, five were from the USA, four were from Canada, and two each were from Belgium, Finland, and England (Table 7).

Table 7: The 20 most productive authors of research papers related to forest ecology indexed using the Web of Science during 2002–2011

	Authors	Authors' countries	Articles number	Ratio (%)
1	Bergeron Y	Canada	146	0.19
2	Kulmala M	Finland	123	0.16
3	Hermy M	Belgium	114	0.14
4	Lindenmayer DB	Australia	110	0.14
5	Black TA	Canada	103	0.13
6	Coops NC	Canada	95	0.12
7	Asner GP	USA	91	0.12
8	Verheyen K	Belgium	91	0.12
9	Reich PB	USA	87	0.11
10	Penuelas J	Spain	85	0.11
11	Vesala T	Finland	85	0.11
12	Leuschner C	Germany	81	0.10
13	Peres CA	England	81	0.10
14	Chen JM	Canada	80	0.10
15	Ciais P	France	80	0.10
16	Groffman PM	USA	79	0.10
17	Law BE	USA	78	0.10
18	Malhi Y	England	78	0.10
19	Fahey TJ	USA	77	0.10
20	Yu GR	China	77	0.10

By Year

From 2002 to 2011, the annual number of published articles about forest ecology was growing at a stable rate (Table 8), as the fit produced a high determination coefficient from the collected data ($R^2 = 0.9955$). The best fit for forest ecology was found to be: $y = 629.75x - 1.2557exp+06$, where y is the article number and x is the number of years since 2002. Extrapolating from the model, the number of articles about forest ecology in the following years could be forecasted (Figure 2).

Table 8: Annual number of articles on forest ecology indexed using the Web of Science during 2002–2011

	Years	Articles number	Ratio (%)
1	2002	5245	6.64
2	2003	5729	7.25

3	2004	6250	7.91
4	2005	6816	8.63
5	2006	7555	9.57
6	2007	8098	10.25
7	2008	8970	11.36
8	2009	9311	11.79
9	2010	10096	12.78
10	2011	10915	13.82

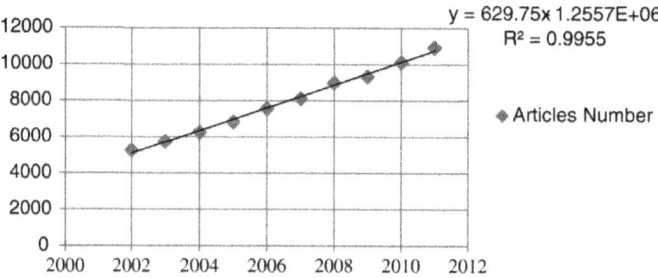

Figure 2: A linear relationship between articles number and years during 2002-2011.

By Publication

The number of journals publishing forest ecology related articles each year increased from 430 in 2002 to 856 in 2011. Table 9 shows the top 20 major journals indicating that *Forest Ecology and Management* (3,876, 4.91%) was the top journal on forest ecology by article count, followed by *Canadian Journal of Forest Research* (1,399, 1.77%) and *Biological Conservation* (1,399, 1.77%).

Table 9: The top 20 journals related to forest ecology analyzed using the Web of Science during 2002–2011

	Publications	Articles number	Ratio (%)
1	Forest Ecology and Management	3876	4.91
2	Canadian Journal of Forest Research	1399	1.77
3	Biological Conservation	933	1.18
4	Soil Biology Biochemistry	929	1.18
5	Biodiversity and Conservation	928	1.18
6	Global Change Biology	824	1.04

7	Ecology	750	0.95
8	Oecologia	741	0.94
9	Biotropica	666	0.84
10	Plant and Soil	653	0.83
11	Ecological Applications	636	0.81
12	Plant Ecology	614	0.78
13	Ecological Modeling	599	0.76
14	Remote Sensing of Environment	598	0.76
15	Argicultural and Forest Meteorology	589	0.75
16	Journal of Tropical Ecology	543	0.69
17	Journal of Geophysical Research Atmospheres	523	0.66
18	Conservation Biology	516	0.65
19	Journal of Biogeography	510	0.65
20	Tree Physiology	508	0.64

DISCUSSION

The results of this study pointed to several significant hotspots in global research related to forest ecology based on an analysis of article keywords for articles published during 2002–2011, and revealed the distribution of the articles from seven aspects listed above. The keyword analysis method and the java analysis program could be extended to other related research fields.

In the keywords analysis, we presumed that a keyword appeared only once in the keywords list of an article (Campbell1963). Therefore the frequency of a keyword could show the number of articles that had used this keyword. For example, the frequency of "forest" was 9,302, meaning that 9,302 articles had used "forest" as a keyword in 73,740 articles.

It was undisputed that "forest" was the most frequently used keyword (9,302 articles). Most writers used this word to express the concept of "forest" instead of its plural "forests"; therefore, "forest" appeared in articles three times more than "forests" (3,069). The next four most frequently used words were "diversity" (5,424), "conservation" (5,135), "dynamics" (4,886), and "vegetation" (4,720) indicating forest diversity, forest conservation, forest dynamics and forest vegetation were the focus of forest ecological studies.

The frequency of "patterns" (4,166), "model" (2,100), and "models" (988) demonstrated that these words were widely used in forest developmental pattern and model studies. The keywords "management" (3,236), "ecology" (2,677), "ecosystems" (2,407), and "ecosystem" (1,362) were also frequently used in macro research (9,682 times), accounting for 1.03% in all keywords indicating large numbers of studies had been carried out in these aspects of forest research in last ten years.

USA" (2,916), "Brazil" (1,018), "Australia" (868), "Mexico" (819), "Costa Rica" (813) and "New Zealand" (796) appeared more frequently than the names of other countries showing that many studies focused on those countries. During the early twenty-first century, the warm droughts in the United States, Europe and Australia have been recognized as a considerable change from the climatological conditions and variability of the late twentieth century (Dai2011), and the focus of forest ecology studies in those regions were impacted accordingly. From a regional point of view, we can see that the total frequencies of "rain-forest" (3,253), "tropical forests" (1,107), and "tropical forest" (1,513) were 5,873, 2.5 times more frequent than "boreal forest" (2,334), indicating that forest ecology studies concerning tropical forests were produced more frequently than those related to boreal forests.

In 2005, large-scale, warm droughts occurred in North America, Africa, Europe, Amazonia and Australia, resulting in major effects on terrestrial ecosystems, carbon balance and food security (Breshears2005). The words "nitrogen" (3,136), "carbon" (2,568), and "phosphorus" (971) were used frequently in the studies concerning elemental nutrients. There were numerous studies related to how the climate is affecting forest ecology, as indicated by the frequencies of "climate-change", "climate", and "climate change," which were 2,412, 2,095 and 1,599, respectively.

This study did reveal some problem areas. Some keywords were not being used consistently, such as soil, soils, forest soil and forest soils, which all pointed to the same thing: forest soil. Another example was that tropical forest and tropical forests also expressed similar meanings. The use of multiple keywords for a single concept might be related to the writing styles and habits of different authors, but this creates difficulty in statistical analysis.

The USA, Canada, and Germany were the top three most productive countries of forest ecology related research. The most three productive organizations were the University of California System, Chinese Academy of Sciences, and the US Forest Service. The three most productive funding agencies were the National Science Foundation, the National Natural Science Foundation of China, and the Natural Sciences and Engineering Research Council of Canada. Environmental science / ecology, forestry, and agriculture were the top three most popular categories. The spatial clusters of authors were mainly in the USA and Canada. *Forest Ecology and Management, Canadian Journal of Forest Research*, and *Biological Conservation* were the top three journals with the most publications related to forest ecology research. In the article analysis, the results by country/region, organization, funding agency, author distribution, and sources titles, was clustered in developed countries, apparently because these countries have economic strength required to invest in science and technology.

In this study, the limitations of search term expressions and the English language made it impossible to include all related keywords in the field of forest ecology research, especially in other languages. This study did not analyze the effects of cooperation between authors and joint papers by authors from multiple nations. In the journal sort, the impact factor of the journal was not considered.

CONCLUSIONS

A serial java program was developed and applied to conduct keyword frequency analysis. That improved the efficiency of data processing and provided an analysis method. Keyword analysis offered insight into forest ecology research areas of interest, while the abundance of less frequent keywords suggested a lack of continuity in research and a wide disparity in the focus of forest ecology research. The top 100 keywords in the keyword analysis were almost all included in the top 20 research areas in the article analysis, so one could conclude that keyword frequency analysis is consistent with article research area analysis. Their difference is the former is concrete and the latter is abstract.

APPENDIX A

(TS = (habitats) or TS = (species diversity) or TS = (biodiversity) or TS = (species richness) or TS = (environmental factors) or TS = (ecosystems) or TS = (plant communities) or TS = (landscape) or TS = (phenology) or TS = (environmental degradation) or TS = (plant) or TS = (populations) or TS = (animal) or TS = (ecological disturbance) or TS = (landscape) or TS = (synecology) or TS = (palaeo ecology) or TS = (community) or TS = (biogeography) or TS = (population) or TS = (ecotypes) or TS = (predator prey relationships) or TS = (microbial) or TS = (freshwater) or TS = (food webs) or TS = (lowland areas) or TS = (restoration) or TS = (fire) or TS = (food chains) or TS = (autecology) or TS = (marine) or TS = (chemical) or TS = (human) or TS = (bioenergetics) or TS = (ecological balance) or TS = (bio coenosis) or TS = (microenvironments) or TS = (dendro ecology) or TS = (climate) or TS = (soils) or TS = (physiography) or TS = (carbon balance) or TS = (nutrient cycling) and (TS = (forest).

ACKNOWLEDGEMENTS

This study was supported by the Forestry Commonweal Programs (No. 200904003) from State Forestry Administration, P.R.China. The authors greatly appreciate the technical support of Peng Shi, Yingni Zou, and Ying Pan. The authors are grateful to Yungang Liao for his helpful suggestions. The authors would also like to thank the chief editor of *SpringerPlus* and anonymous reviewers for their valuable comments.

AUTHORS' CONTRIBUTIONS

YS carried out the bibliometric analysis, and drafted the manuscript. TZ gave financial assistance and some advice to this manuscript. Both authors read and approved the final manuscript.

REFERENCES

1. Barnes BV, Zak DR, Denton SR, Spurr SH: *Forest ecology*. 1997. http://www.cabdirect.org/abstracts/19980607188.html
2. Berger TW, Inselsbacher E, Mutsch F, Pfeffer M: Nutrient cycling and soil leaching in eighteen pure and mixed stands of beech (Fagus sylvatica) and spruce (Picea abies). *Forest Ecol Manag* 2009, 258(11):2578-2592. 10.1016/j.foreco.2009.09.014
3. Breshears DD: Regional vegetation die-off in response to global-change-type drought. *Proc Natl Acad Sci* 2005, 102: 15144-15148. 10.1073/pnas.0505734102
4. Campbell DJ: Making your own indexing system in science and technology (classification and keyword systems). *ASLIB Proc* 1963., 15(10):
5. Cheaib A, Badeau V, Boe J, Chuine I, Delire C, Dufrêne E: Climate change impacts on tree ranges: model intercomparison facilitates understanding and quantification of uncertainty. *Ecol Lett* 2012, 15(6):533-544. 10.1111/j.1461-0248.2012.01764.x
6. Dai A: Drought under global warming: a review. *WIRES Climate Change* 2011, (2):45-65.
7. Diaz S, Lavorel S, de Bello F, Quetier F, Grigulis K, Robson M: Incorporating plant functional diversity effects in ecosystem service assessments. *Proc Natl Acad Sci USA* 2007, 104(52):20684-20689. 10.1073/pnas.0704716104
8. Dobbertin M, Nobis M: Exploring research issues in selected forest journals 1979–2008. *Ann Forest Sci* 2010, 67(8):800. 10.1051/

forest/2010052

9. Grandjean P, Eriksen M, Ellegaard O, Wallin J: The Matthew effect in environmental science publication: A bibliometric analysis of chemical substances in journal articles. *Environ Health* 2011, 10(1):1. 10.1186/1476-069X-10-1

10. Hanberry BB, Kabrick JM, He HS, Palik BJ: Historical trajectories and restoration strategies for the Mississippi River Alluvial Valley. *Forest Ecol Manag* 2012, 280: 103-111. 10.1016

11. Hendrix D: An analysis of bibliometric indicators, National Institutes of Health funding, and faculty size at Association of American Medical Colleges medical schools, 1997–2007. *J Med Libr Assoc* 2008, 96(4):324-334. 10.3163/1536-5050.96.4.007

12. Lamb D, Erskine P, Parrotta J: Restoration of degraded tropical forest landscapes. *Science* 2005, 310(5754):1628-1632. 10.1126/science.1111773

13. Loucks CJ, Lu Z, Dinerstein E, Wang H, Olson DM, Zhu C: Ecology. Giant pandas in a changing landscape. *Science* 2001, 294(5546):1465. 10.1126/science.1064710

14. McLachlan SM, Bazely DR: Outcomes of longterm deciduous forest restoration in southwestern Ontario, Canada. *Biol Conserv* 2003, 113(2):159-169. 10.1016/S0006-3207(02)00248-3

15. Mitchell SR, Harmon ME, O'Connell KEB: Forest fuel reduction alters fire severity and long-term carbon storage in three Pacific Northwest ecosystems. *Ecol Appl* 2009, 19(3):643-655. 10.1890/08-0501.1

16. Morrissey RC, Gauthier M, Kershaw JA Jr, Jacobs DF, Seifert JR, Fischer BC: Grapevine (Vitis spp.) dynamics in association with manual tending, physiography, and host tree associations in temperate deciduous forests. *Forest Ecol Manag* 2009, 257(8):1839-1846. 10.1016/j.foreco.2009.02.004

17. Narotsky D, Green PH, Lebwohl B: Temporal and geographic trends in celiac disease publications: a bibliometric analysis. *Eur J Gastroenterol Hepatol* 2012, 24(9):1071-1077. 10.1097/MEG.0b013e328355a4ab

18. Perez M, Fu M, Xie J, Yang X, Belcher B: The relationship between forest research and forest management in China: an analysis of four leading Chinese forestry journals. *Int For Rev* 2004, 6(3–4):341-345.

19. Rodrigues RR, Gandolfi S, Nave AG, Aronson J, Barreto TE, Vidal CY: Large-scale ecological restoration of high-diversity tropical forests in SE Brazil. *Forest Ecology Manag* 2011, 261(10):1605-1613. 10.1016/j.

foreco.2010.07.005

20. Rubio A, Escudero A: Effect of climate and physiography on occurrence and intensity of decarbonation in Mediterranean forest soils of Spain. *Geoderma* 2005, 125(3–4):309-319.

21. Şekercioğlu ÇH, Primack RB, Wormworth J: The effects of climate change on tropical birds. *Biol Conserv* 2012, 148(1):1-18. 10.1016/j.biocon.2011.10.019

22. Sillett SC, Van Pelt R, Koch GW, Ambrose AR, Carroll AL, Antoine ME: Increasing wood production through old age in tall trees. *Forest Ecol Manag* 2010, 259(5):976-994. 10.1016/j.foreco.2009.12.003

23. Sitzia T, Semenzato P, Trentanovi G: Natural reforestation is changing spatial patterns of rural mountain and hill landscapes: A global overview. *Forest Ecol Manag* 2010, 259(8):1354-1362. 10.1016/j.foreco.2010.01.048

24. van Eck NJ, Waltman L, Dekker R, van den Berg J: A comparison of two techniques for bibliometric mapping: Multidimensional scaling and VOS. *J Am Soc Inf Sci Technol* 2010, 61(12):2405-2416. 10.1002/asi.21421

25. van Oudenhoven APE, Petz K, Alkemade R, Hein L, de Groot RS: Framework for systematic indicator selection to assess effects of land management on ecosystem services. *Ecol Indic* 2012, 21: 110-122.

26. van Raan AFJ: Comparison of the Hirsch-index with standard bibliometric indicators and with peer judgment for 147 chemistry research groups. *Scientometrics* 2006, 67(3):491.

27. Wang Q, Wang S, Yu X: Decline of soil fertility during forest conversion of secondary forest to Chinese fir plantations in substropical China.*Land Degradation Dev* 2011, 22(4):444-452. 10.1002/ldr.1030

28. Wintle B, Bekessy S, Venier L, Pearce J, Chisholm R: Utility of dynamic-landscape metapopulation models for sustainable forest management. *Conserv Biol* 2005, 19(6):1930-1943. 10.1111/j.1523-1739.2005.00276.x

29. Xu Z, Chen C: Fingerprinting global climate change and forest management within rhizosphere carbon and nutrient cycling processes. *Environ Sci Pollut Res* 2006, 13(5):293. 10.1065/espr2006.08.340

Chapter 12

SIZE-ENERGY RELATIONSHIPS IN ECOLOGICAL COMMUNITIES

Brent J. Sewall[1,2], Amy L. Freestone[1], Joseph E. Hawes[3], Ernest Andriamanarina[4]

[1] Department of Biology, Temple University, Philadelphia, Pennsylvania, United States of America

[2] Department of Wildlife, Fish, and Conservation Biology, University of California Davis, Davis, California, United States of America

[3] School of Environmental Sciences, University of East Anglia, Norwich Research Park, Norwich, United Kingdom

[4] De´partement des Sciences de la Nature et de l'Environnement, Universite´ d'Antsiranana, Antsiranana, Madagascar

ABSTRACT

Hypotheses that relate body size to energy use are of particular interest in community ecology and macroecology because of their potential to facilitate quantitative predictions about species interactions and to clarify complex ecological patterns. One prominent size-energy hypothesis, the energetic equivalence hypothesis, proposes that energy use from shared, limiting resources by populations or size classes of foragers will be independent of body size. Alternative hypotheses propose that energy use will increase with body size, decrease with body size, or peak at an intermediate body size. Despite extensive study, however, size-energy hypotheses remain controversial, due to a lack of directly-measured data on energy use, a tendency to confound distinct scaling relationships, and insufficient attention to the ecological contexts in which predicted relationships are likely to occur. Our goal, therefore, was to directly evaluate size-energy hypotheses while clarifying how results would differ with alternate methods and assumptions. We comprehensively tested size-energy hypotheses in a vertebrate frugivore guild in a tropical forest in Madagascar. Our test of size-energy hypotheses, which is the first to examine energy intake directly, was consistent with the energetic equivalence hypothesis. This finding corresponds with predictions of metabolic theory and models of energy distribution in ecological communities, which imply that body size does not confer an advantage in competition for energy among populations or size classes of foragers. This result was robust to different assumptions about energy regulation. Our results from direct energy measurement, however, contrasted with those obtained with conventional methods of indirect inference from size-density relationships, suggesting that size-density relationships do not provide an appropriate proxy for size-energy relationships as has commonly been assumed. Our research also provides insights into mechanisms underlying local size-energy relationships and has important implications for predicting species interactions and for understanding the structure and dynamics of ecological communities.

INTRODUCTION

Body size has strong, widespread, and predictable relationships to organisms' physiological traits and life history characteristics [1], [2], and has therefore been proposed as a primary driver of diverse patterns in community ecology and macroecology [2], [3]. Relationships between body size and energy use in particular have long intrigued ecologists [4], [5], because they promise to reveal an underlying order in highly complex ecological systems, and to predict quantitative patterns of species interactions from local to global spatial scales.

One prominent hypothesis relating body size to energy use was proposed by Damuth, who focused on three variables of interest – the density (D) of a population, the mean metabolic (respiration) rate (R) of individuals of that population, and the total amount of energy used by all individuals of that population, or population energy use (PEU) – as well as the relationships of each of these three variables to body size, as indicated by body mass (M). In a widely-cited paper [6], he suggested that population energy use could be estimated as the product of population density and mean metabolic rate (i.e., $PEU = D * R$). To relate these variables to body size, he compiled a global dataset of population densities of mammalian herbivores from published accounts, and found that population density (D) scaled globally with body mass (M) to the power of -0.75 [6]. Noting that metabolic rate (R) was already understood to scale with body mass (M) to the power of $+0.75$ [1], Damuth then hypothesized that the total energy used by a population should generally be independent of body size [6], because $PEU = D * R \propto M^{-0.75} * M^{0.75} = M^0$. This finding, later called the 'energetic equivalence rule' [7], has been highly influential, because it suggests that underlying the complexities of ecological communities, populations of animals of all body sizes are on equal footing in the competition for energy (i.e., there is no optimal body size where energy use is maximized) [6]. The energetic equivalence rule has also garnered widespread attention since it provides a quantifiable prediction for a point of great interest in ecology – the distribution of energy among populations in a community – on the basis of a relatively easily-measured variable, body size. This energetic equivalence rule, and its emphasis on three-quarter-power scaling with body mass, has provided the theoretical foundation for explaining a range of ecological patterns, from population growth [8] to community structure [9] to global biodiversity patterns [10].

Later authors [5], [11], [12] expanded on this idea, noting that population energy use would also be independent of body size wherever both scaling exponents were exactly inversely correlated. Thus, if $D \propto M^a$, $R \propto M^b$, and $PEU \propto M^c$, then $M^a * M^b = M^c$ and population energy use will be independent of body size whenever $c = 0$. This will occur whenever $-a = b$, even if $-a$ and b are not equal to 0.75 in the system being studied. Thus, the 'energetic equivalence rule' is a special case (where $a = -0.75$, $b = +0.75$) of a broader hypothesis known as 'energetic equivalence' [5], where $-a = b$. This energetic equivalence hypothesis predicts that inverse scaling relationships of population density and metabolic rate are widespread, and that population energy use by different species is broadly independent of body size in ecological communities [13].

While the energetic equivalence hypothesis has been supported in some studies [14], [15], this hypothesis is only one possible expectation for energy use within ecological communities [4]. Other studies have proposed and provided evidence in support of alternative hypotheses: that energy use increases with body size (e.g., [16]), decreases with body size (e.g., [12]), or peaks at an intermediate body size (e.g., [17]). Thus, despite intense interest in size-energy hypotheses, and despite their potential importance for understanding community structure and dynamics, past studies have produced conflicting results and the influence of body size on energy use in ecological communities remains unclear.

In addition, size-energy hypotheses have been controversial because they are often disconnected from the proposed mechanism underlying these hypotheses – namely, that size-energy relationships result from energetic tradeoffs due to resource competition among organisms of different body sizes in a community [2], [5], [6] but see [18]. This mechanism implies that size-energy hypotheses apply only in certain ecological contexts – namely those where all species can directly compete for shared, limiting resources. Previous studies of size-energy hypotheses, however, have tended to examine large spatial scales by compiling data from many different communities around the globe (e.g., [6]), though species selected for analysis from separate communities cannot plausibly compete [19]. Further, those studies that have examined the local scale have not controlled for resource availability (e.g., [20]), though the use of different resources implies a lack of energetic tradeoffs due to competition [2], [20]. Inferences about size-energy hypotheses from large spatial scale or multi-resource studies should therefore be interpreted carefully.

Size-energy hypotheses, and past tests of these hypotheses, have also been controversial on methodological grounds, in part due to a tendency to confound distinct scaling relationships in two primary ways. First, all previous studies have indirectly used size-density relationships, rather than size-energy relationships, to test size-energy hypotheses (e.g., [11], [21], [22]). This method (hereafter, the 'indirect method') has spread because in practice it is easier to measure the density scaling relationship (i.e., the exponent a) than the energy scaling relationship (c) in a community. In such studies, the exponent of the metabolic scaling relationship (b) is assumed to be a universal constant (+0.75) (e.g., [6], [14], [15], [23]), or is taken from published metabolic scaling relationships when available (e.g., [24]), such that c can be inferred directly from a via the equation $M^a * M^b = M^c$, or $a+b = c$. However, recent study suggests that metabolic scaling exponents are highly variable [25], and so a specific exponent cannot be assumed for all systems (i.e., b is not constant across all systems, and thus a is insufficient to determine c). This approach also suffers from a lack of data on species-specific metabolic scaling exponents and is subject to the hidden non-linearity, propagation of error variances, and inherent imprecision of multiplying allometric relationships [21], [26]. This practice is also circular, since in the analysis, body size serves both as the predictor variable (M) and as a means to calculate the response variable (PEU); results therefore will tend to be biased to suggest a positive correlation. As a result, although size-density relationships are interesting in their own right [5], and can provide insight into mechanisms underlying energy use, their utility for estimating the form of the size-energy relationships that are the focus of size-energy hypotheses or for providing a robust test of these hypotheses remains unclear [19]. Direct measurement of energy use (hereafter, the 'direct method') could avoid these problems, but measurement of energy use by all species in a community is technically challenging [4], and no test of size-energy hypotheses using direct measurements of energy use has previously been attempted [27].

Second, two distinct approaches to evaluating size-energy hypotheses have emerged on the basis of the underlying assumptions for how energy use is regulated in ecological communities. Competition for energy implies zero-sum dynamics [13], [28], where limited resources are allocated to individual organisms in the community [29] and thus increases in energy consumed from a shared, limiting resource by one individual are necessarily offset by equal and opposite decreases in energy consumed by other individuals. Authors of studies on terrestrial animals have generally assumed that such competition for energy is regulated by competition among species populations in a community (e.g., [6]), while authors of studies on trees and marine systems have generally assumed that energy use is regulated by competition among size classes in a community (e.g., [30], [31]). For the former assumption, population energy use (*PEU*) is used to evaluate the size-energy relationship, whereas for the latter assumption, a related but distinct variable – size class energy use (*SCEU*) – is used. Unlike *PEU*, which is calculated as the sum of the energy used by all individuals in a species population, *SCEU* is calculated as the sum of the energy used by all individuals in a given size class, regardless of species identity. Because size classes may vary in the number of species they contain, and individuals of a single species may be distributed among multiple size classes, tests of size-energy hypotheses using each approach may provide differing results for the same community[32]. Despite these differences, reviews of size-energy hypotheses have typically not distinguished between the population and size-class approaches, leading to confusion in the literature [5]. Comparison of the two approaches in the same system could clarify underlying assumptions of how energy use is regulated in communities. Such comparisons have been completed for size-density relationships in multi-trophic communities [32] (see also [5], [33] fordiscussion), but not for size-energy relationships.

Our goal in this project was to directly evaluate size-energy hypotheses while clarifying how results would differ with alternate methods and assumptions. We had three objectives in support of this goal. First, we sought to complete the first analyses of size-energy hypotheses to directly measure how energy use scales with body size. Specifically, we evaluated whether energy use is independent of body size, increases with body size, decreases with body size, or peaks at an intermediate body size. Second, we sought to evaluate the influence of indirect versus direct methods on conclusions about size-energy hypotheses. Specifically, we compared results of our novel direct method with those from the indirect method conventionally used to test size-energy hypotheses. Third, we sought to determine the influence of assumptions about energy regulation on conclusions about size-energy hypotheses. Specifically, we compared results obtained using the population and size-class approaches. We addressed these objectives by examining scaling relationships among a guild of vertebrate frugivores as they foraged on a common set of shared, limiting resources in a tropical forest in Madagascar.

MATERIALS AND METHODS

Ethics Statement

This study was conducted at all times in strict accordance with the animal welfare protocols of the University of California, Davis and with the laws of the participating countries. Animal research was non-manipulative and completely observational, and solely examined wild animals in their natural habitat. The study was conducted on protected land owned by the Madagascar government and managed by Madagascar National Parks, and was approved by the Madagascar Ministère de l'Environnement, Direction Générale des Eaux et Forêts, under permits N° 143/MINEV.EF/SG/DGEF/DPB/SCBLF/RECH, N° 225/06/MINEV.EF/SG/DGEF/DPB/SCBLF/RECH, N° 024N-EV01/MG06, and N° 308N-EV11/MG06.

Study System

We focused on frugivores, or fruit eating animals, due to their central ecological importance for the maintenance and regeneration of tropical forests [34], [35]. Specifically, we studied the vertebrate frugivore guild of Ankarana National Park, near Mahamasina in northern Madagascar, within 175 ha of semi-evergreen primary forest. This region undergoes two distinct seasons: a short wet season from January–April when almost all of the ~1890 mm of annual rainfall occurs, and a longer dry season during the remaining months ([36]; pers. obs.). The study period was the end of the dry season (October–December 2005 and October 2006-early January 2007), at a time when available fruit is in limited supply and competition for shared resources is expected to be greatest. Frugivores are abundant and easily observable within the park, where they directly compete for clearly identifiable resources – primarily fruit from the canopies of fruiting trees. The frugivore guild comprised five bird, five primate (lemur), and two fruit bat species, a taxonomically diverse fauna with distinct foraging behaviors and body sizes spanning two orders of magnitude (Table 1).

Table 1. Vertebrate frugivore guild of Ankarana National Park, Madagascar, with taxa, activity, mass, and size class

Species	English name	Taxon	Activity	Mass (g)[a]	Size class
Saroglossa aurata	Madagascar Starling	Bird	Diurnal	40	A
Hypsipetes madagascariensis	Madagascar Bulbul	Bird	Diurnal	43.5	A
Microcebus tavaratra[b,c]	northern rufous mouse lemur	Lemur	Nocturnal	64.5	B
Rousettus madagascariensis	Madagascar rousette	Bat	Nocturnal	65	B
Cheirogaleus medius[c]	fat-tailed dwarf lemur	Lemur	Nocturnal	198	C
Treron australis	Madagascar Green Pigeon	Bird	Diurnal	236	C
Coracopsis nigra	Lesser Vasa Parrot	Bird	Diurnal	254	D
Eidolon dupreanum	Madagascar straw-colored fruit bat	Bat	Nocturnal	295	D
Phaner electromontis	Amber Mountain fork-marked lemur	Lemur	Nocturnal	425	E
Coracopsis vasa	Greater Vasa Parrot	Bird	Diurnal	530	E
Eulemur coronatus	crowned lemur	Lemur	Cathemeral	1450	F
Eulemur sanfordi	Sanford's brown lemur	Lemur	Cathemeral	2150	G

[a] Based on the midpoint of the body mass range of adults of each species [72,73,74].
[b] We considered all *Microcebus* observations at Ankarana to be of *M. tavaratra* on the basis of recent taxonomic analyses [75].
[c] These or related species may enter torpor seasonally at other sites [73], but we observed them foraging throughout our study period.
doi:10.1371/journal.pone.0068657.t001

Indirect method for estimating size-energy relationships from size-density relationships

We first used the conventional indirect method for testing size-energy hypotheses, by measuring size-density relationships using frugivore population densities. For this portion of the study, we focused on frugivorous lemurs and bats. To determine lemur population densities, we established 12 standardized line transects of 500 m in length in primary forest, ≥100 m from each other and the forest edge. We walked each transect three times, during peak periods of frugivore foraging just after sunrise (05:00–08:00), and just before (15:00–18:00) and just after sunset (18:30–21:30). To control for animal behavior and detectability, we did not conduct observations during rain or wind, or during moonlit nights. We walked transects at 10 m per minute, searching both sides of the transect line for visual or auditory signs of lemurs. During nocturnal observations, we used two observers to ensure adequate coverage [37], and headlamps to facilitate lemur detections via direct sighting and eyeshine [38]. Upon detection, we identified each lemur to species and estimated perpendicular distance from the line to lemur individuals. To calculate densities of lemur populations from transect data, we determined a standard suite of detection functions separately for each species. We controlled for different effort during diurnal and nocturnal sampling with a multiplier [39], and we calculated densities of the two cathemeral species, *Eulemur coronatus* and *E. sanfordi*, solely from diurnal periods, when they were most active. We selected the best detection models following well-established methodology for distance sampling analyses [39], [40], and using Distance 6.0.2 software [41].

Since fruit bats were not easily detectable with distance sampling methods, we used roost count data to estimate their population densities. Because both species, *Rousettus madagascariensis* and *Eidolon dupreanum*, are capable of flying long distances, we used published estimates of their abundance at all roosts near Ankarana National Park [42], divided by the area of all primary and secondary forest in the same region [43].

We then converted densities to energy use with the conventional indirect approach, multiplying population density by individual field metabolic rate to determine energy use per unit area (kJ/(hr * km^2)). As in past studies, we determined metabolic rates from allometric scaling relationships of similar species in the literature (equation 2 for eutherian mammals in [24]).

Direct Method for Measuring Size-Energy Relationships

We went beyond this conventional indirect method by developing a direct method for measuring the size-energy relationship. Size-energy hypotheses are typically formulated in relation to 'energy use' (e.g., population energy use, *PEU*), the amount of energy processed by all organisms of a population during metabolism, but this metric cannot easily be measured directly in the field. However, 'energy intake', the amount of metabolizable energy consumed by an organism, is both directly observable and closely correlated with energy use [44]. Energy intake is also directly linked to the proposed mechanism (of resource competition) that underlies size-energy hypotheses [5], [6], since energy consumed by one organism from a shared, limited resource in a habitat is not available to another organism, regardless of whether that consumed energy is actually processed. We therefore used energy intake as our metric by measuring fruit consumption during both diurnal and nocturnal foraging periods by the entire frugivore guild – including frugivorous birds, lemurs, and bats.

To determine the size-energy relationship, we measured energy intake by frugivore species while foraging at fig (*Ficus*, Moraceae) trees as a proxy for total energy intake. This approach assumes that figs were accessible to all frugivores and that the relative portion of energy intake at fig trees by frugivores was proportional to their total energy intake during the study period. The general importance of the frugivore-fig interaction, and several characteristics of our site – including strong preferences by frugivores for fig fruit, the abundance of fig fruit, scarcity of other resources, and evidence of rapid fig consumption – together suggest this assumption is justified in our study system. Specifically, the frugivore-fig interaction is critical to tropical forest communities as figs are widely distributed [45] and common in tropical forests [46]. Figs also play a keystone role in maintaining frugivore populations [47], [48], as they provide non-defended, easy-to-digest syconia (hereafter, fruit) year-round that is critical to a wide variety of frugivores during periods of fruit scarcity [48]. At our site, observations of frugivore foraging in fig trees and in the eight other most common non-fig species of fruiting trees during the dry season indicated strong preferences by all frugivore species for fig fruit, and no body-size trend among frugivores in relative preferences for fig fruit (KE Reuter, AA Gudiel, S Nieves, C Stanley, BJ Sewall, unpublished). During our study period in the late dry season, non-fig fruits were scarce: on the basis of fruit crop measurements at fig trees, and fruit counts at 10 randomly-placed transects of 50 m×2 m observed at two-week intervals, the three most common fig species provided a mean of 62.6% of all ripe fruits (range: 45–84%) in the habitat. During this period, fig fruit was highly prized: frugivores completely depleted even the largest fig fruit crops (up to ~80,000 fruits) within one month after the onset of ripening, and often within 1–2 weeks. Thus fig fruits represent a shared, limited resource for the entire frugivore guild. Of the six fig tree species in the study area, we focused on the three most common: *Ficus grevei*, *F. polita* (formerly *F. megapoda*), and *F. reflexa*, which together comprised 81% of all fig tree individuals, and 95% of all fig trees to reach peak fruiting during the study period.

We determined energy intake from fruit consumption via direct, unobtrusive observations on free-ranging frugivores in the field at focal fig trees. We observed foraging by frugivores during peak fruiting stages at each of 34 trees (18 *F. grevei*, seven *F. polita*, and nine *F. reflexa*). At each tree, we conducted observations during peak periods of both diurnal and nocturnal foraging. We used focal observations of individual frugivores at each tree to determine the fruit consumption rate (the rate at which an individual of a frugivore species consumed fig fruit) for each frugivore species at each fig species. At each tree we also used repeated scan censuses of the tree canopy to determine the residency rate (the number of minutes spent in the tree's canopy by all individuals of a species per hour) for each frugivore species at each tree. For each fig species, we used laboratory analysis of collected fruit samples to determine the metabolizable energy available per fruit. We then multiplied the fruit consumption rate (fruits/individual-min), the residency rate (individual-min/hr), and the metabolizable energy per fruit (kJ/fruit) to determine the energy intake rate by each frugivore species at each tree (kJ/hr). We then summed across all fruiting figs in the habitat and converted to energy intake per unit area (kJ/(hr * km^2)). Further details of these data collection protocols are in Appendix S1.

Analyses

To test size-energy hypotheses, we determined size-energy relationships with both the indirect and direct methods, and with both the population and size-class approaches, rendering four size-energy scaling relationships. All variables were natural log transformed prior to analysis. For the population approach, population density or population energy intake was regressed onto the body mass of each species. For the size-class approach, we assigned each frugivore species to a size class on the basis of adult body mass (Table 1), and we then summed density or energy intake across all species in the size class. We used normalized-logarithmic binning to assess the form of the size-class relationships. We chose normalized-logarithmic binning because linear binning is known to provide inaccurate estimates of parameters [49]. We did not use maximum likelihood estimation because we expected exponents >-1 and methods for handling these exponents are not commonly available and require estimating a maximum size, which is unknown. Size classes were 0.5 natural log unit bins, normalized for width of the bins (following [49]). Size-class density or size-class energy intake was then regressed onto the midpoint of each size-class bin.

For each of the four size-energy scaling relationships, we first compared the hypothesized linear relationships (energetic equivalence, energy intake increases with body size, energy intake decreases with body size) with a quadratic model for the hypothesized non-linear relationship (energy intake peaks at an intermediate body size). The relative suitability of the linear and quadratic models given the data was evaluated with a model estimation and selection process using information theory (following [50]). While such a model selection process is suitable for the population approach, the process is an ad hoc method for model selection in the size-class approach (see [51] for discussion of issues related to model selection in size-class models). Conclusions would not have differed had we not used a model selection procedure and instead simply used the full model, due to low support for the quadratic term. Since linear models were always selected via the selection procedure (see Results), we then determined the slope (scaling exponent) of the relationship for the selected models from the log-log plot. Because the sign of the slope of these relationships was not known *a priori* and under some hypotheses was expected to be zero, it was not appropriate to use standardized major axis regression [52]. Therefore, we determined slopes in the conventional manner, with ordinary least squares regression (following [6], [14], [16]). To evaluate correlations between the indirect and direct methods, we compared energy use values for all species or size classes common to each pair of size-energy relationships with Pearson's correlation coefficient. Finally, to understand factors that may affect correlations among the size-energy relationships, we used the mathematical links between scaling exponents ($a+b = c$; see Introduction) in the population approach to calculate the size-metabolism exponent b from the measured size-density exponent a and the measured size-energy exponent c. We completed all statistical analyses with JMP 8.0 [53] and R 2.11.1 [54] statistical software.

RESULTS

Indirect Method for Estimating Size-Energy Relationships from Size-Density Relationships

We detected 389 frugivorous lemurs in transects. Overall, frugivorous mammal densities ranged from 65.2 individuals per km² in *E. dupreanum* to 189.7 individuals per km² in *E. coronatus* (Appendix S2). Based on these data, the linear models for the size-energy relationship with the indirect method received more support than the quadratic models (Table 2). The slope of the size-density relationship using the population approach ($a = -0.11$, $F_{1,5} = 23.23$, $p = 0.005$; Fig. 1A) was shallower (less negative) than expected under the energetic equivalence rule (where a is hypothesized to equal -0.75). Evidence for a difference from the energetic equivalence rule was weaker in the size-density relationship, however, when this relationship was examined with the size-class approach ($a = -0.40$, $F_{1,4} = 2.60$, $p = 0.18$; Fig. 1B). The slope of the size-energy relationship calculated with the indirect method was greater than expected under energetic equivalence (where c is hypothesized to equal 0) with the population approach ($c = 0.66$, $F_{1,5} = 24.85$, $p = 0.004$; Fig. 1C), suggesting that results from the indirect method support the hypothesis that population energy use increases linearly with body size. Evidence for such a difference from energetic equivalence in the size-energy relationship with the indirect method, however, was weaker when this relationship was examined with the size-class approach ($c = 0.37$, $F_{1,4} = 2.91$, $p = 0.16$; Fig. 1D).

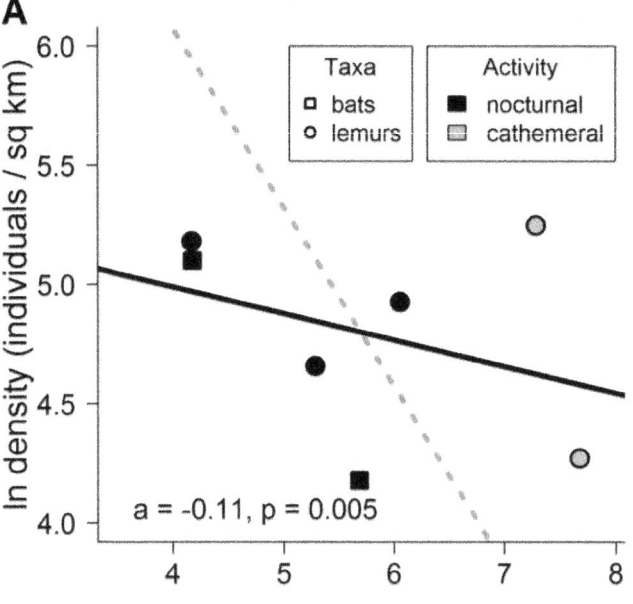
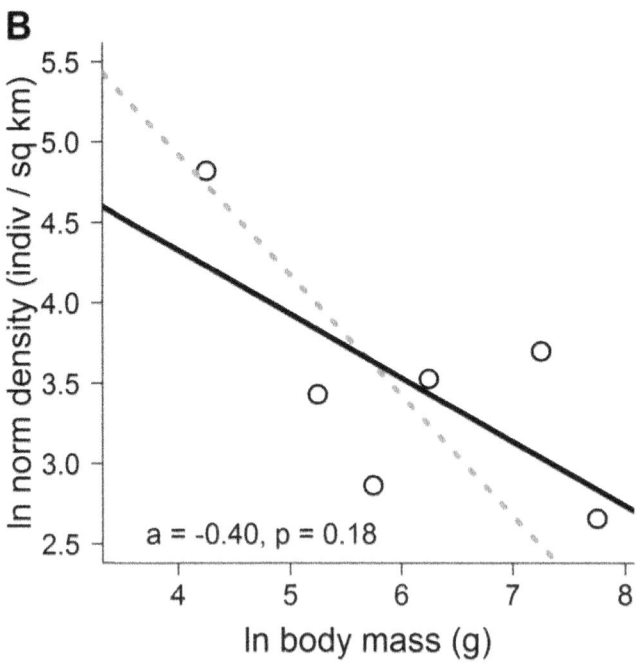

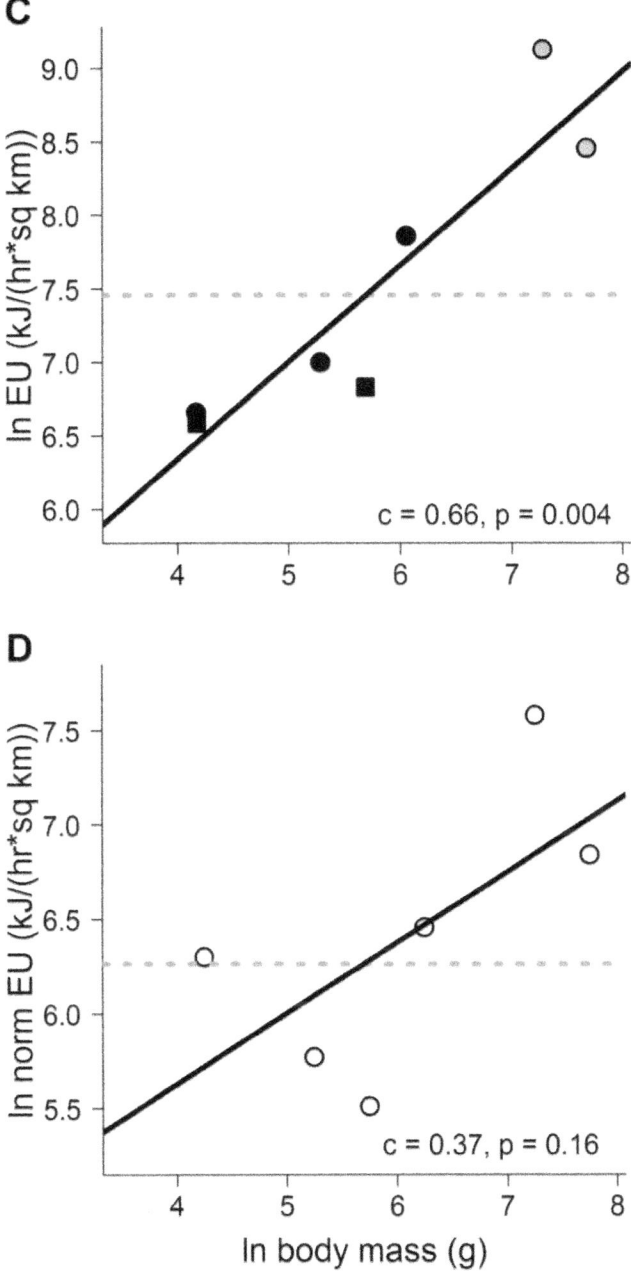

Figure 1. Size-energy relationships, as determined from the indirect method of inference from size-density relationships. (A) The regression slope (solid line) of the size-density relationship with the population approach was significantly shallower (less negative) than the expected slope under the energetic equivalence rule ($a = -0.75$, dot-

ted line). (B) Evidence for a difference from the energetic equivalence rule was weaker in the size-density relationship with the size-class approach. (C) When calculated with the indirect method, the size-energy relationship (where the size-energy scaling exponent, c, was equal to 0.66) with the population approach was significantly more positive than expected under energetic equivalence ($c = 0$, dotted line). (D) Evidence was weaker for a difference from energetic equivalence for the size-energy relationship when examined with the size-class approach ($c = 0.37$). Note that EU = energy use, and the y-axes in the size-class analyses (B and D) are normalized for bin width.

Table 2. Comparison of linear and quadratic models for log-log regressions of size-energy relationships, examined with both the indirect and direct methods and both the population and size-class approaches

Method	Approach	Model Type	AICc	Δ_i	w_i
Indirect	Population	Linear	20.21	0	0.999
		Quadratic	33.46	13.26	0.001
	Size class	Linear	27.12	0	1.000
		Quadratic	55.18	28.06	0.000
Direct	Population	Linear	63.26	0	0.861
		Quadratic	66.92	3.66	0.139
	Size class	Linear	37.19	0	0.999
		Quadratic	50.65	13.46	0.001

Best models were selected on the basis of the small sample size corrected Akaike's Information Criterion (AICc); Δ_i is the difference in AICc between the best and the alternate model, and w_i is the Akaike weight, the weight of evidence for each model given the data (where 1.000 represents the highest likelihood of the model relative to the alternate model).
doi:10.1371/journal.pone.0068657.t002

Direct method for measuring size-energy relationships

At our 34 focal trees, we conducted 272 person-hours of observations, with 4080 scan sample observations and 241 focal observations of frugivore foraging of at least 30 seconds in duration. We also collected 1159 fruits from nine fig trees for fruit content analysis (Appendix S1). Population energy intake ranged from 0.25 kJ/(hr * km²) by *Phaner electromontis* to 937 kJ/(hr * km²) by *Treron australis* (Appendix S2). Based on these data, the linear models for the size-energy relationship with the direct method received more support than the quadratic models (Table 2). Using the direct method, the slopes of the size-energy relationship using both the population approach ($c = 0.31$, $F_{1,10} = 0.28$, p

= 0.61; Fig. 2A) and the size-class approach ($c = -0.08$, $F_{1,5} = 0.036$, $p = 0.86$; Fig. 2B) were not different from zero, suggesting that results from the direct method were consistent with the energetic equivalence hypothesis.

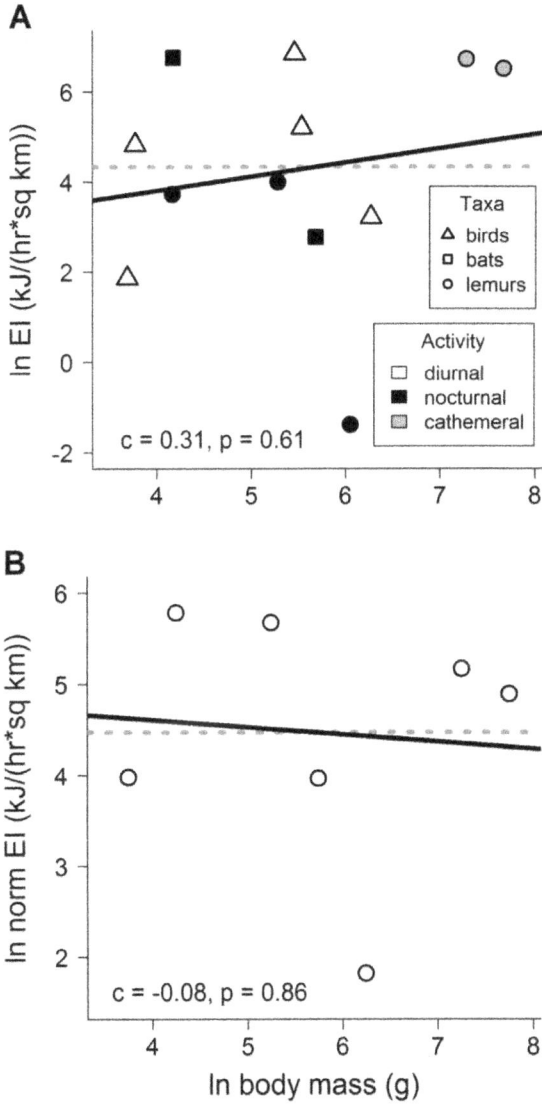

Figure 2. Size-energy relationships, as determined from the direct method of measuring energy intake. Regression slopes (solid lines) of the size-energy relationship with the (A) population approach (where the size-energy scaling exponent, c, was equal to 0.31), and (B) size-class appr($c = -0.08$) were both not significantly different from zero, and therefore were both consistent with the expected slope under energetic

equivalence ($c = 0$, dotted lines). Slopes were not qualitatively different if potential outliers (i.e., low values in A and B) were removed. Note that EI = energy intake, and the y-axis in the size class analysis (B) is normalized for bin width.

Correlations among Size-Energy Relationships and Implications for Size-Metabolism Relationships

The indirect and direct methods of estimating size-energy relationships were uncorrelated (population approach, $r = 0.19$, $p = 0.68$; size-class approach, $r = 0.07$, $p = 0.89$). Finally, in the population approach, the slopes of our measured size-density relationship (Fig. 1A) and our measured size-energy relationship (Fig. 2A) imply the slope of the size-metabolism relationship b is 0.42.

DISCUSSION

Size-Energy Hypotheses—Direct Method

Size-energy hypotheses promise to clarify complex patterns of resource use in ecological communities and to provide quantitative predictions of species interactions, but such hypotheses have remained controversial. Rigorous empirical evaluation of size-energy hypotheses requires direct measurement of energy use by foragers during competition for shared, limiting resources [2], [19], but until now tests using direct measurements of energy use had not been completed. We conducted the first such direct test of size-energy hypotheses, and our results (Table 2, Fig. 2) were consistent with the central prediction of the energetic equivalence hypothesis, namely that energy use does not vary systematically with body size. These findings correspond with predictions of metabolic theory [2] and models of energy distribution in ecological communities [55], which suggest a general principle for community ecology: that body size does not confer an advantage to populations or size classes in the competition for energy from shared, limiting resources in communities.

Size-Energy Hypotheses—Comparison of Direct and Indirect Methods

For comparison with previous studies, we also estimated energy use with the conventional indirect method of inference from size-density relationships and assumptions of constant allometric relationships with metabolism. Previous studies of size-energy hypotheses, which have all used this method, have produced variable results, with most community-scale studies indicating that energy use peaks at intermediate body sizes [13], [17], [20], [26], [30] or increases with body size [11], [16], [21], [22], [56]. These results have been invoked to explain a number of broad ecological patterns, including the island rule (the common pattern of dwarfism and gigantism on isolated islands; [17]), Cope's Rule (the common evolutionary cycle in which large-bodied organisms emerge and become extinct more rapidly than small-bodied species;[16]), and the distribution of colony sizes in eusocial species [22].

Our results from the indirect method and population approach indicated that the slope of the size-energy relationship was linear (Table 2) and positive ($c>0$; Fig. 1C), suggesting that populations of animals of large body size dominate others in the competition for energy. These results were similar to many previous community-scale studies [11], [16], [21], [22], [56], yet these results contrasted with our directly-measured results from the same community (Fig. 2). The indirectly-calculated slope with the size-class approach was also linear (Table 2) but did not differ significantly from the expectation of energetic equivalence (Fig. 1D). Finally, in all cases, results from the indirect method were uncorrelated with results from the direct method (all $r \leq 0.19$, all $p \geq 0.68$, see Results).

The discrepancy in results from the direct and indirect methods – at least in the population approach – may occur due to the inherent biases and assumptions of the indirect method. Notably, the direction of the difference (relatively greater energy use at larger body sizes with the indirect method) represents just the kind of positive correlation between body size and energy use that is expected to derive from the use of body size as a component of both the predictor and response variable in the conventional method of determining and analyzing size-energy relationships (see Introduction).

Alternatively, this discrepancy could also reflect the slightly different focus of both methods. Our study focused on long-lived species, and thus the indirect method, due to its reliance on population densities to estimate energy use, may integrate energy use over all resources and over multiple years. In contrast, our direct method targeted key food resources during a critical period of the year for frugivore survival. Mismatches between the two methods could arise if organisms respond differently to temporal variation in food abundance on the basis of body size, such as for example, if small and large organisms differ in ability (e.g., due to different home range sizes or search areas) to access patchy and unpredictable fruit resources during periods of fruit scarcity [57]. Thus, differences in size-energy relationships resulting from the indirect and direct methods may represent size-dependent temporal variation in energy use.

The contrasting results from the two methods suggest size-density relationships are not appropriate proxies for size-energy relationships. This is not to argue against the examination of size-density relationships, which have provided important insights into key patterns of broad significance for ecology and macroecology [5]. A variety of other promising approaches are also emerging to address energy use in communities [28], [29]. However, our results highlight the limitations and challenges of applying density measurements to understand size-energy relationships, and call into question theories for size-energy relationships that have been derived from measured densities rather than energy use. Without further efforts to empirically validate or refute hypothesized relationships between the two variables with independent lines of evidence, the utility of the density-based methods for estimating local energy scaling relationships will remain unclear.

Processes of Energy Regulation—Comparison of Population and Size-Class Assumptions

Different versions of size-energy hypotheses imply alternative assumptions about processes of energy regulation in communities, and analyses of size-density relationships in multi-trophic communities indicate that energy regulation assumptions may strongly influence observed scaling patterns [2], [23], [32]. In size-energy relationships, population and size-class approaches assume that energy use is driven by metabolic processes related to an organism's taxonomic identity or its body size, respectively [2]. We provided the first empirical comparison of these alternative assumptions for size-energy relationships in the same system, through the use of both population and size-class approaches to quantifying scaling relationships. Although size structure was skewed in our guild toward species with smaller body sizes (Table 1), as is typical of many guilds [58], directly-measured results from the population approach (Fig. 2A) and the size-class approach (Fig. 2B) nonetheless were not different, suggesting that our conclusions about size-energy hypotheses are robust to the energy regulation assumption used. Future comparisons of energy-regulation assumptions could benefit from new techniques for the size-class approach that account for intraspecific variation in body size during species sorting into size class bins [59].

Mechanisms Underlying Size-Energy Relationships

Theory suggests that the energy use of populations or size classes is dependent on their density and individual metabolic rate [2], [6]. While results from our measured size-energy relationship are consistent with energetic equivalence (Fig. 2), results from our measured size-density relationship (Fig. 1A, 1B) add to accumulating evidence [5], [21] that the slope of the size-density relationship is shallower (less negative) at local scales than the −0.75 slope expected from global-scale studies [60], metabolic theory [2], and the energetic equivalence rule [6]. The combination of these two sets of results is unexpected because the conventional indirect method assumes the slope of the size-metabolism relationship, b, is +0.75, a value which corresponds with global patterns in size-metabolism relationships [1], [6], [24]. However, our combined results can be explained if the slope of the local size-metabolism relationship in our system was shallower (less positive) than expected. Specifically, our results for the population approach imply a slope of the local size-metabolism relationship of +0.42 (see Results). Local-scale patterns of size-metabolism relationships for interacting species within a community have not previously been explored, but such differences in local-scale variation from global-scale patterns could be common, since metabolic rate is widely variable among taxa [25] and with activity level [61]. In our system the shallower-than-expected slope of the size-metabolism relationship may result from the relatively slow metabolism of frugivores and arboreal species [62], [63] or the energetic constraints imposed by food-scarcity at the end of the dry season [48], [62].

Challenges in Testing Energetic Equivalence

In addition to the methodological and analytical challenges discussed above, several other factors may complicate tests of scaling relationships in general and the energetic equivalence hypothesis in particular. First, statistical methods for scaling relationships have been controversial, with particular debate over means to reduce bias in estimation of scaling exponents (e.g, [12], [64]). Statistical testing of the energetic equivalence hypothesis poses an additional challenge, as the hypothesis posits a slope of the relationship between body size and energy use of zero, yet regression, the statistical test used to evaluate such relationships, is not designed to enable detection of a zero slope but rather to distinguish deviations from zero [65]. Energetic equivalence is treated statistically as a null hypothesis, and although non-significant results such as those we obtained for the direct measurement of energy use (Fig. 2) have often been taken as positive evidence of the invariance of energy use with body size (e.g., [14], [15]), they cannot be distinguished statistically from a failure to detect a difference from zero [27]. Further application of statistical methods to address these issues is needed [27], [64].

Second, tests of energetic equivalence often exhibit substantial variation around mean trends (e.g., [15]), and both simulation models [66] and empirical evidence [67] predict higher variation at smaller body size ranges. Such variation may decrease power and make it harder to detect scaling relationships empirically in community-scale studies [27]. Greater power could be obtained by examining a greater number of species or a greater body mass range, but this will not be possible to accomplish while controlling for resource use in many communities such as ours, where species numbers and body mass ranges in a guild are limited (Table 1; [66]).

In our study, data were highly variable for all relationships (Figs. 1, 2), but variance around the mean scaling trend was much greater for the direct method (standard deviation >2 natural log units; Fig. 2A, 2B) than for the indirect method (st. dev. <0.5 natural log units; Figs. 1C, 1D). This discrepancy in variances in the two methods may be due to the use of an assumed constant metabolic scaling exponent, b, of +0.75 for all species to calculate energy use in the conventional indirect method (see Introduction), despite the highly variable nature of this exponent by taxon [25] or community (as noted above, b differed from this assumption for our system). By ignoring such community-scale variation in size-metabolism relationships, the conventional indirect approach may obscure an important source of error and greatly overestimate the precision of reported size-energy relationships. Any future analyses using the indirect method should explicitly account for this expected variance in the metabolic scaling exponent.

While size-energy relationships offer an explanation for the mean scaling trend among species in a community, they do not provide an explanation for the high levels of residual variation in the data. Such variation may be due to differences in a number of evolutionary and ecological factors, including relatedness, niche differences, dispersal patterns, predation risk, or trade-offs among growth, mortality, fecundity and other life history variables [28], [29], [66], [68]. Understanding the relative influence of body size and each of these other factors on energy use remains an important question in community ecology [29]. Further insights can be gained through studies comparing the relative influence of each variable on energy use, and through additional data collected across communities and over time [27], [28].

CONCLUSIONS

Investigations of local scaling relationships enable prediction of community patterns of population density and resource use; provide important links among individual-, population-, and community-level processes in ecology; and clarify mechanisms underlying resource partition in ecological communities [5]. By providing the first direct measurement of the size-energy relationship, and by controlling for factors that have often confounded previous studies, our study enabled new insights into the form of the local size-energy relationship. Specifically, our study is consistent with the idea that energy use is independent of body size. Our results were robust to the energy regulation assumption used, and suggest that shallower-than-expected relationships of body size to density and metabolism may combine to produce energetic equivalence at local scales. Our study focused on foragers on shared, limiting resources in communities because competition for resources is the proposed mechanism underlying energetic equivalence, but such results can be extended to multitrophic communities [2], [23], [32], [64], [69], [70].

In addition to directly evaluating the size-energy relationship, our study compared alternative methods to understand this relationship at the community scale. Our direct method of measurement of the size-energy relationship contrasts sharply with interpretations that would have been derived from the conventional indirect method, suggesting that the size-density relationship in a community cannot be used to infer the size-energy relationship. Further, the differences we observed in size-density relationships between our local study and past global studies highlight that body size relationships may describe distinct phenomena, be generated by differing mechanisms, and exhibit contrasting patterns at different spatial scales.

More broadly, our focus in this study was on evaluating size-energy hypotheses, and our finding that energy intake was independent of body size (Fig. 2) was consistent with the energetic equivalence hypothesis. This finding is of particular interest for ecological theory, since energetic equivalence implies that small and large animals will produce similar amounts of biomass over time [6]; population growth and carrying capacities will be related predictably to body size [8]; resources will be used at similar rates by groups of organisms of different sizes[2]; and community assembly will not be driven by or oriented around an optimal body size [71]. Our findings therefore have important implications for predicting species interactions and for understanding many aspects of community structure and dynamics.

ACKNOWLEDGMENTS

We thank Madagascar National Parks, the Universities of Antsiranana and Antananarivo, Madagascar Institut pour la Conservation des Ecosystèmes Tropicaux, Madagascar CAFF/CORE, Direction des Eaux et Forêts, Missouri Botanical Garden, and the villagers of Mahamasina for cooperation; J. Jamildine, M. Jaoravo, J. Robert, P. Randrianantenaina, R. Andrianjaka, S. Rakotondriabe, G. Papan'i Dada, and K. Arison for field assistance; A. Abdel Cader, L. Miadana, and the Ankarana staff and guides for logistical support; D. Van Vuren for advice and support; and D. Van Vuren, D. Kelt, A. Sih, R. Greenberg, N. Isaac, E. White, and an anonymous reviewer for comments on previous versions of the manuscript.

AUTHOR CONTRIBUTIONS

Analyzed the data: BJS. Wrote the paper: BJS. Conceived of project: BJS. Developed conceptual framework: BJS ALF. Designed observations: BJS ALF JEH EA. Managed field teams: BJS JEH. Collected data: BJS ALF JEH EA. Edited the manuscript: BJS ALF JEH.

REFERENCES

1. Kleiber M (1975) The Fire of Life, 2nd edition. New York: Krieger.
2. Brown JH, Gillooly JF, Allen AP, Savage VM, West GB (2004) Toward a metabolic theory of ecology. Ecology 85: 1771–1789. doi: 10.1890/03-9000
3. Brown JH, Maurer BA (1989) Macroecology: the division of food and space among species on continents. Science 243: 1145–1150. doi: 10.1126/science.243.4895.1145
4. Blackburn TM, Gaston KJ (1996) On being the right size: different definitions of 'right'. Oikos 75: 551–557. doi: 10.2307/3545899
5. White EP, Ernest SKM, Kerkhoff AJ, Enquist BJ (2007) Relationships between body size and abundance in ecology. Trends in Ecology & Evolution 22: 323–330. doi: 10.1016/j.tree.2007.03.007
6. Damuth J (1981) Population-density and body size in mammals. Nature 290: 699–700. doi: 10.1038/290699a0
7. Nee S, Read AF, Greenwood JJD, Harvey PH (1991) The relationship between abundance and body size in British birds. Nature 351: 312–313. doi: 10.1038/351312a0
8. Savage VM, Gillooly JF, Brown JH, West GB, Charnov EL (2004) Effects of body size and temperature on population growth. American Naturalist 163: 429–441. doi: 10.1086/381872
9. Enquist BJ, West GB, Brown JH (2009) Extensions and evaluations of a general quantitative theory of forest structure and dynamics. Proceedings of the National Academy of Sciences of the United States of America 106: 7046–7051. doi: 10.1073/pnas.0812303106
10. Allen AP, Brown JH, Gillooly JF (2002) Global biodiversity, biochemical kinetics, and the energetic-equivalence rule. Science 297: 1545–1548. doi: 10.1126/science.1072380
11. Russo SE, Robinson SK, Terborgh J (2003) Size-abundance relationships in an Amazonian bird community: implications for the energetic equivalence rule. American Naturalist 161: 267–283. doi: 10.1086/345938

12. Griffiths D (1992) Size, abundance, and energy use in communities. Journal of Animal Ecology 61: 307–315. doi: 10.2307/5323
13. Marquet PA, Quinones RA, Abades S, Labra F, Tognelli M, et al. (2005) Scaling and power-laws in ecological systems. Journal of Experimental Biology 208: 1749–1769. doi: 10.1242/jeb.01588
14. Enquist BJ, Brown JH, West GB (1998) Allometric scaling of plant energetics and population density. Nature 395: 163–165.
15. Ernest SKM, Enquist BJ, Brown JH, Charnov EL, Gillooly JF, et al. (2003) Thermodynamic and metabolic effects on the scaling of production and population energy use. Ecology Letters 6: 990–995. doi: 10.1046/j.1461-0248.2003.00526.x
16. Brown JH, Maurer BA (1986) Body size, ecological dominance and Cope's rule. Nature 324: 248–250. doi: 10.1038/324248a0
17. Damuth J (1993) Cope's rule, the island rule and the scaling of mammalian population density. Nature 365: 748–750. doi: 10.1038/365748a0
18. DeLong JP, Vasseur DA (2012) A dynamic explanation of size-density scaling in carnivores. Ecology 93: 470–476. doi: 10.1890/11-1138.1
19. Loeuille N, Loreau M (2006) Evolution of body size in food webs: does the energetic equivalence rule hold? Ecology Letters 9: 171–178. doi: 10.1111/j.1461-0248.2005.00861.x
20. Ernest SKM (2005) Body size, energy use, and community structure of small mammals. Ecology 86: 1407–1413. doi: 10.1890/03-3179
21. Hayward A, Khalid M, Kolasa J (2009) Population energy use scales positively with body size in natural aquatic microcosms. Global Ecology and Biogeography 18: 553–562. doi: 10.1111/j.1466-8238.2009.00459.x
22. King JR (2010) Size-abundance relationships in Florida ant communities reveal how ants break the energetic equivalence rule. Ecological Entomology 35: 287–298. doi: 10.1111/j.1365-2311.2010.01177.x
23. Brown JH, Gillooly JF (2003) Ecological food webs: high-quality data facilitate theoretical unification. Proceedings of the National Academy of Sciences of the United States of America 100: 1467–1468. doi: 10.1073/pnas.0630310100
24. Nagy KA, Girard IA, Brown TK (1999) Energetics of free-ranging mammals, reptiles, and birds. Annual Review of Nutrition 19: 247–277. doi: 10.1146/annurev.nutr.19.1.247
25. Isaac NJB, Carbone C (2010) Why are metabolic scaling exponents so controversial? Quantifying variance and testing hypotheses. Ecology Letters 13: 728–735. doi: 10.1111/j.1461-0248.2010.01461.x

26. Marquet PA, Navarrete SA, Castilla JC (1995) Body size, population density, and the Energetic Equivalence Rule. Journal of Animal Ecology 64: 325–332. doi: 10.2307/5894

27. Isaac NJB, Storch D, Carbone C (2013) The paradox of energy equivalence. Global Ecology and Biogeography 22: 1–5. doi: 10.1111/j.1466-8238.2012.00782.x

28. Ernest SKM, Brown JH, Thibault KM, White EP, Goheen JR (2008) Zero sum, the niche, and metacommunities: long-term dynamics of community assembly. American Naturalist 172: E257–E269. doi: 10.1086/592402

29. Isaac NJB, Carbone C, McGill BJ (2012) Population and community ecology. In: Sibly RM, Brown JH, Kodric-Brown A, editors. Metabolic Ecology: A Scaling Approach. West Sussex, U.K.: John Wiley & Sons. pp. 77–85.

30. Ackerman JL, Bellwood DR, Brown JH (2004) The contribution of small individuals to density-body size relationships: examination of energetic equivalence in reef fishes. Oecologia 139: 568–571. doi: 10.1007/s00442-004-1536-0

31. West GB, Enquist BJ, Brown JH (2009) A general quantitative theory of forest structure and dynamics. Proceedings of the National Academy of Sciences of the United States of America 106: 7040–7045. doi: 10.1073/pnas.0812294106

32. Reuman DC, Mulder C, Raffaelli D, Cohen JE (2008) Three allometric relations of population density to body mass: theoretical integration and empirical tests in 149 food webs. Ecology Letters 11: 1216–1228. doi: 10.1111/j.1461-0248.2008.01236.x

33. Yvon-Durocher G, Reiss J, Blanchard J, Ebenman B, Perkins DM, et al. (2011) Across ecosystem comparisons of size structure: methods, approaches and prospects. Oikos 120: 550–563. doi: 10.1111/j.1600-0706.2010.18863.x

34. Cordeiro NJ, Ndangalasi HJ, McEntee JP, Howe HF (2009) Disperser limitation and recruitment of an endemic African tree in a fragmented landscape. Ecology 90: 1030–1041. doi: 10.1890/07-1208.1

35. Nunez-Iturri G, Howe HF (2007) Bushmeat and the fate of trees with seeds dispersed by large primates in a lowland rain forest in western Amazonia. Biotropica 39: 348–354. doi: 10.1111/j.1744-7429.2007.00276.x

36. Hawkins AFA, Chapman P, Ganzhorn JU, Bloxam QMC, Barlow SC, et al. (1990) Vertebrate conservation in Ankarana Special Reserve, northern Madagascar. Biological Conservation 54: 83–110. doi: 10.1016/0006-3207(90)90136-d

37. Nekaris KAI, Blackham GV, Nijman V (2008) Conservation implications of low encounter rates of five nocturnal primate species (*Nycticebus* spp.) in Asia. Biodiversity and Conservation 17: 733–747. doi: 10.1007/s10531-007-9308-x
38. Radhakrishna S, Goswami AB, Sinha A (2006) Distribution and conservation of *Nycticebus bengalensis* in northeastern India. International Journal of Primatology 27: 971–982. doi: 10.1007/s10764-006-9057-9
39. Buckland ST, Anderson DR, Burnham KP, Laake JL, Borchers DL, et al.. (2001) Introduction to Distance Sampling: Estimating Abundance of Biological Populations. New York: Oxford University Press.
40. Thomas L, Buckland ST, Rexstad EA, Laake JL, Strindberg S, et al. (2010) Distance software: design and analysis of distance sampling surveys for estimating population size. Journal of Applied Ecology 47: 5–14. doi: 10.1111/j.1365-2664.2009.01737.x
41. Thomas L, Laake JL, Rexstad E, Strindberg S, Marques FFC, et al.. (2010) Distance 6.0. Release 2: Research Unit for Wildlife Population Assessment, University of St. Andrews, UK. Program Distance website. Available: http://www.ruwpa.st-and.ac.uk/distance/. Accessed 2010 Aug 13.
42. Cardiff SG, Ratrimomanarivo FH, Rembert G, Goodman SM (2009) Hunting, disturbance and roost persistence of bats in caves at Ankarana, northern Madagascar. African Journal of Ecology 47: 640–649. doi: 10.1111/j.1365-2028.2008.01015.x
43. Fowler SV, Chapman P, Checkley D, Hurd S, McHale M, et al. (1989) Survey and management proposals for a tropical deciduous forest reserve at Ankarana in northern Madagascar. Biological Conservation 47: 297–313. doi: 10.1016/0006-3207(89)90072-4
44. Clauss M, Schwarm A, Ortmann S, Streich WJ, Hummel J (2007) A case of non-scaling in mammalian physiology? Body size, digestive capacity, food intake, and ingesta passage in mammalian herbivores. Comparative Biochemistry and Physiology A-Molecular & Integrative Physiology 148: 249–265. doi: 10.1016/j.cbpa.2007.05.024
45. Berg CC (1989) Classification and distribution of *Ficus*. Experientia 45: 605–611. doi: 10.1007/bf01975677
46. Janzen DH (1979) How to be a fig. Annual Review of Ecology and Systematics 10: 13–51. doi: 10.1146/annurev.es.10.110179.000305
47. Shanahan M, So S, Compton SG, Corlett R (2001) Fig-eating by vertebrate frugivores: a global review. Biological Reviews 76: 529–572. doi: 10.1017/s1464793101005760

48. Terborgh J (1986) Keystone plant resources in the tropical forest. In: Soule ME, editor. Conservation Biology: The Science of Scarcity and Diversity. Sutherland, Massachusetts: Sinauer Associates. pp. 330–344.

49. White EP, Enquist BJ, Green JL (2008) On estimating the exponent of power-law frequency distributions. Ecology 89: 905–912. doi: 10.1890/07-1288.1

50. Burnham KP, Anderson DR (2002) Model Selection and Multimodel Inference: A Practical Information-Theoretic Approach. 2nd edition. New York: Springer. 488 p.

51. Edwards AM, Freeman MP, Breed GA, Jonsen ID (2012) Incorrect likelihood methods were used to infer scaling laws of marine predator search behaviour. PLOS One 7. doi: 10.1371/journal.pone.0045174

52. Warton DI, Wright IJ, Falster DS, Westoby M (2006) Bivariate line-fitting methods for allometry. Biological Reviews 81: 259–291. doi: 10.1017/s1464793106007007

53. SAS Institute (2008) JMP. Version 8.0. Statistical software 1989–2008. Cary, NC.

54. R Development Core Team (2010) R Version 2.11.1. R Foundation for Statistical Computing. Vienna, Austria. R Project website. Available: http://www.R-project.org. Accessed 2010 July 20.

55. Damuth J (2007) A macroevolutionary explanation for energy equivalence in the scaling of body size and population density. American Naturalist 169: 621–631. doi: 10.1086/513495

56. Arneberg P, Skorping A, Read AF (1998) Parasite abundance, body size, life histories, and the energetic equivalence rule. American Naturalist 151: 497–513. doi: 10.1086/286136

57. Saracco JF, Collazo JA, Groom MJ (2004) How do frugivores track resources? Insights from spatial analyses of bird foraging in a tropical forest. Oecologia 139: 235–245. doi: 10.1007/s00442-004-1493-7

58. Blackburn TM, Gaston KJ (1994) Animal body size distributions: patterns, mechanisms, and implications. Trends in Ecology & Evolution 9: 471–474. doi: 10.1016/0169-5347(94)90311-5

59. Thibault KM, White EP, Hurlbert AH, Ernest SKM (2011) Multimodality in the individual size distributions of bird communities. Global Ecology and Biogeography 20: 145–153. doi: 10.1111/j.1466-8238.2010.00576.x

60. Damuth J (1987) Interspecific allometry of population density in mammals and other animals: the independence of body mass and population energy use. Biological Journal of the Linnean Society 31: 193–246. doi: 10.1111/j.1095-8312.1987.tb01990.x

61. Nagy KA (2005) Field metabolic rate and body size. Journal of Experimental Biology 208: 1621–1625. doi: 10.1242/jeb.01553
62. McNab BK (1986) The influence of food habits on the energetics of eutherian mammals. Ecological Monographs 56: 1–19. doi: 10.2307/2937268
63. McNab BK (2003) Metabolism - Ecology shapes bird bioenergetics. Nature 426: 620–621. doi: 10.1038/426620b
64. Arim M, Berazategui M, Barreneche JM, Ziegler L, Zarucki M, et al.. (2011) Determinants of Density-Body Size Scaling Within Food Webs and Tools for Their Detection. In: Belgrano A, Reiss J, editors. Advances in Ecological Research, Vol 45: The Role of Body Size in Multispecies Systems. pp. 1–39.
65. Fisher RA (1925) Statistical Methods for Research Workers, 14th edition. Reprinted in 1990 in Statistical Methods, Experimental Design and Scientific Inference. Oxford, U.K.: Oxford University Press.
66. Tilman D, Hillerislambers J, Harpole S, Dybzinski R, Fargione J, et al. (2004) Does metabolic theory apply to community ecology? It's a matter of scale. Ecology 85: 1797–1799. doi: 10.1890/03-0725
67. Hayward A, Kolasa J, Stone JR (2010) The scale-dependence of population density-body mass allometry: Statistical artefact or biological mechanism? Ecological Complexity 7: 115–124. doi: 10.1016/j.ecocom.2009.08.005
68. Cotgreave P (1993) The relationship between body size and population abundance in animals. Trends in Ecology & Evolution 8: 244–248. doi: 10.1016/0169-5347(93)90199-y
69. Hechinger RF, Lafferty KD, Dobson AP, Brown JH, Kuris AM (2011) A common scaling rule for abundance, energetics, and production of parasitic and free-living species. Science 333: 445–448. doi: 10.1126/science.1204337
70. Reuman DC, Mulder C, Banasek-Richter C, Blandenier MFC, Breure AM, et al.. (2009) Allometry of Body Size and Abundance in 166 Food Webs. In: Caswell H, editor. Advances in Ecological Research, Vol 41. pp. 1–44.
71. Kelt DA (1997) Assembly of local communities: consequences of an optimal body size for the organization of competitively structured communities. Biological Journal of the Linnean Society 62: 15–37. doi: 10.1111/j.1095-8312.1997.tb01615.x
72. Ravokatra M, Wilmé L, Goodman SM (2003) Bird weights. In: Goodman SM, Benstead JP, editors. The Natural History of Madagascar. Chicago: The University of Chicago Press. pp. 1059–1063.

73. Mittermeier RA, Konstant WR, Hawkins AFA, Louis EE, Langrand O, et al.. (2006) Lemurs of Madagascar. Washington, D.C.: Conservation International. 520 p.
74. Garbutt N (2007) Mammals of Madagascar: A Complete Guide. New Haven, Connecticut: Yale University Press.
75. Rasoloarison RM, Goodman SM, Ganzhorn JU (2000) Taxonomic revision of mouse lemurs (*Microcebus*) in the western portions of Madagascar. International Journal of Primatology 21: 963–1019.

Chapter 13

SEISMIC MICROZONATION OF BREGINJSKI KOT (NW SLOVENIA) BASED ON DETAILED ENGINEERING GEOLOGICAL MAPPING

Jure Kokošin[1] and Andrej Gosar[2,3]

[1] Sweco Norge AS, Dronningensgate 52/54, 8509 Narvik, Norway

[2] Faculty of Natural Sciences and Engineering, University of Ljubljana, Aškerčeva 12, SI-1000 Ljubljana, Slovenia

[3] Slovenian Environment Agency, Seismology and Geology Office, Dunajska 47, SI-1000 Ljubljana, Slovenia

ABSTRACT

Breginjski kot is among the most endangered seismic zones in Slovenia with the seismic hazard assessed to intensity IX MSK and the design ground acceleration of 0.250 g, both for 500-year return period. The most destructive was the 1976 Friuli Mw = 6.4 earthquake which had maximum intensity VIII-IX. Since the previous microzonation of the area was based solely on the basic geological map and did not include supplementary field research, we have performed a new soil classification of the area. First, a detailed engineering geological mapping in scale 1: 5.000 was conducted. Mapped units were described in detail and some of them interpreted anew. Stiff sites are composed of hard to medium-hard rocks which were subjected to erosion mainly evoked by glacial and postglacial age. At that time a prominent topography was formed and different types of sediments were deposited in valleys by mass flows. A distinction between sediments and weathered rocks, their exact position, and thickness are of significant importance for microzonation. On the basis of geological mapping, a soil classification was carried out according to the Medvedev method (intensity increments) and the Eurocode 8 standard (soil factors) and two microzonation maps were prepared. The bulk of the studied area is covered by soft sediments and nine out of ten settlements are situated on them. The microzonation clearly points out the dependence of damage distribution in the case of 1976 Friuli earthquake to local site effects.

INTRODUCTION

The Breginjski kot (NW Slovenia, Figure 1) is located close to a seismically very active area of Friuli in NE Italy, but also the Julian Alps in Slovenia have recently experienced an increase of seismic activity. In this area, in fact, the seismic hazard in Slovenia is the highest. Most of numerous settlements in the area are built on soft sediments which can significantly enhance the ground shaking. The whole area was highly damaged during the Friuli 1976 Mw = 6.4 earthquake and its aftershock sequence. For a realistic assessment of side effects and earthquake hazard, a thorough seismic microzonation is therefore essential. On the other hand the existing microzonation of Breginjski kot [1] was based solely on the basic geological map of Slovenia in scale 1:100.000, without supplementary field investigations or subsurface information, thus making it fairly inaccurate. Therefore we have decided to perform a new microzonation study based on a detailed engineering geological mapping in scale 1:5.000. Such a map would enable soil classification according to two different approaches without using expensive geotechnical drilling or geophysical investigations.

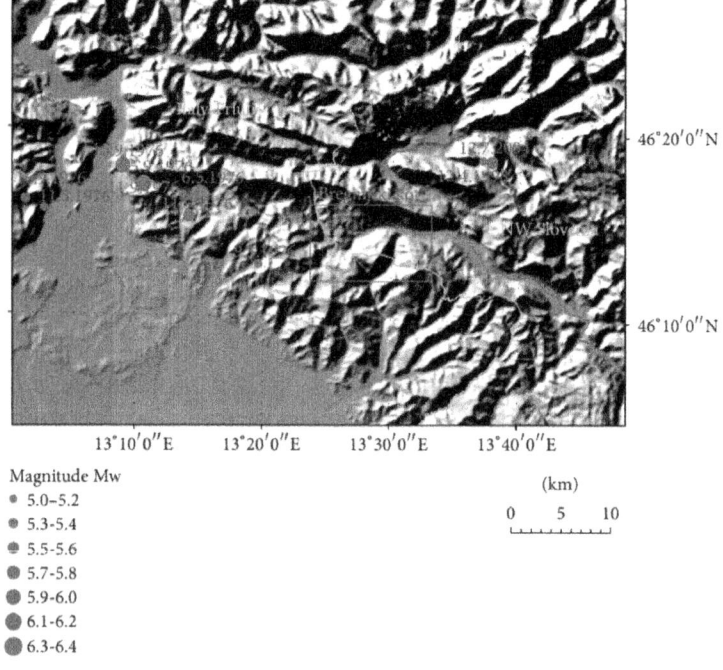

Figure 1: Seismicity (Mw ≥ 5.0) in Friuli and NW Slovenia with location of Breginjski kot study area.

The seismic microzonation according to Eurocode 8 was based on a new European seismic standard [2] which has been applied in Slovenia from 2008. This standard defines the seismic hazard by the peak ground acceleration, whereas the site effects are expressed by the soil factor. The product of both values is used in the design of earthquake resistant buildings. The seismic microzonation according to the Medvedev method [3] denotes seismic hazard by increments of seismic intensity. Although the seismic intensity in Slovenia is not used for building design anymore, it is still important for seismological analyses and for the civil protection purposes.

SEISMOLOGICAL CHARACTERISTICS

The Breginjski kot is located in one of the three areas with the highest seismic hazard in Slovenia. This is mainly due to the proximity of the seismically very active area of Friuli located 10–30 km to the west in NE Italy. In this area the Mw = 6.4 Friuli earthquake occurred in May 1976. The seismic sequence that followed (Table 1, Figure 1) consisted of six events of Mw between 5.0 and 6.0 during the same year and one event of Mw = 5.3 that occurred in September 1977 [4, 5]. The intensity of the main shock in Breginj was VIII EMS-98, but the cumulative intensity which includes also damage from aftershocks reached IX EMS-98. The highest intensity in Slovenia from the main shock was VIII-IX in Robidišče [6]. A complete review of 1976 earthquake intensities in Breginjski kot settlements is summarized in Table 2.

Table 1: Data for earthquakes with Mw ≥ 5.0 in the vicinity of Breginski kot (after [4, 7,8])

Date	Time (UTC)	Lat (°N)	Lon (°N)	Depth (km)	Mw	Region
06/05/1976	20:00	46.275	13.246	6.5	6.4	Friuli
09/05/1976	00:53	46.214	13.326	8.5	5.1	Friuli
11/05/1976	22:44	46.260	13.041	6.0	5.0	Friuli
11/09/1976	16:31	46.275	13.224	6.5	5.2	Friuli
11/09/1976	16:35	46.233	13.233	4.3	5.6	Friuli
15/09/1976	03:15	46.284	13.173	5.0	5.9	Friuli
15/09/1976	09:21	46.300	13.145	7.0	6.0	Friuli
16/09/1977	23:48	46.268	13.016	8.0	5.3	Friuli
12/04/1998	10:55	46.309	13.632	7.6	5.6	Krn Mts.
12/07/2004	13:04	46.310	13.620	11.0	5.2	Krn Mts.

Table 2: Intensities of 1976 earthquake in different settlements of Breginjski kot

Settlement	Intensity EMS-98
Borjana	VI
Kred	VI
Robič	VI
Potoki	VI+ (MSK-78)
Logje	VI-VII
Staro Selo	VII
Sedlo	VII-VIII
Breginj	VIII
Podbela	VIII
Robidišče	VIII-IX

For the NW Slovenia relatively weak rates of seismicity were observed before April 1998 when a Mw = 5.6 earthquake occurred in Krn Mountains, only 7 km NW from Kobarid [9, 10]. It was followed by a Mw = 5.2 event in July 2004 in the same epicentral area (Figure 1). Both earthquakes occurred on the NW-SE trending near-vertical dextral strike-slip Ravne fault [11] in the depth range 7.6–11 km. The intensity of the 1998 earthquake in Kobarid (located at eastern margin of the Breginjski kot) was VI-VII EMS-98 [7] and of the 2004 event VI EMS-98 [8].

NW Slovenia and Friuli region are located at the kinematic transition between E-W striking thrust faults of the Alpine system (Friuli earthquakes) and NW-SE striking strike-slip faults of the Dinarides system (Krn Mountains earthquakes) (Figure 1). The strongest earthquake ever recorded in the Alps-Dinarides junction area was the 1511 western Slovenia earthquake (Mw = 6.8); the exact location and mechanism of this event are still debated [12, 13] due to early occurrence and thus very little or no historical documents.

According to the old seismic hazard map of Slovenia for a 500-year return period showing expected intensities on the MSK scale [14], Breginjski kot is estimated to have IX MSK intensity. According to the new seismic hazard map for a 475-year return period [15], a design ground acceleration value for a rock site in Breginjski kot is 0.250 g. The important methodological difference between both maps is that intensities are assessed for a "medium" soil type, whereas design ground acceleration is assessed for a "rock" site according to Eurocode 8.

Accelerographs were installed in Kobarid after the main Friuli shock in 1976 and after the Krn Mountains earthquake in 1998. The strongest ground motion in the Friuli seismic sequence was recorded for the September 15, 1976 aftershock (Mw = 6.0, distance = 37 km) as 0.138 g peak ground acceleration [16]. For the July 12, 2004 Krn Mountain earthquake (Mw = 5.2, distance = 7 km) a peak ground acceleration of 0.152 g was recorded [17].

The previously existing microzonation of Breginjski kot [1] was based solely on the basic geological map in scale 1: 100.000, without supplementary field research or subsurface geological and geophysical information, thus making it fairly inaccurate. It was prepared to be used together with the old seismic hazard map showing intensities on the MSK scale. This microzonation showed that the maximum expected intensity due to the effects of soft sediments can be increased by up to one intensity degree.

ENGINEERING GEOLOGICAL MAPPING

A detailed engineering geological map of Breginjski kot in scale 1: 5.000 was prepared by outcrop observing and geological boundaries tracking method. Since seismic microzonation was the final goal, the main focus of the field survey was to characterise local geological conditions in vicinity of settlements and other substantial infrastructure like roads, power lines, water collectors, and so forth. Consequently some less relevant lithological boundaries were summarised from the manuscript geological map in scale 1:25.000, which was the basis for published basic geological map in scale 1:100.000 [18]. Based on the mapping accomplished, rocks and sediments were classified by their characteristics determined in the field. Properties such as structure, dip, strength and weathered material thickness were defined for rocks, whereas granulation, roundness, and thickness were described for unconsolidated sediments. Mapped lithological units were compared with the Bovec basin, located 15 km to the NE in the Soča valley for which intensity increment microzonation was performed after the 1998 earthquake [19]. Lithologies described below (Figures 2 and 3) are arranged by age order and are presented on a detailed engineering geological map (Figure 4) of Breginjski kot. Their names and age classification mainly coincide with lithologies described in the explanatory text of the basic geological map of Slovenia [20].

258 Ecology and Applied Environmental Science

Figure 2: Typical lithologies according to the engineering-geological mapping: (a) Megalodontidae shell in Dachstein limestone in Robič, (b) limestone breccia interbedded with shalestone near Robidišče, (c) folds in flyschoid formation near Borjana, (d) heterogeneous composition of diamicton in Sedlo, (e) lacustrine chalk SW of Podbela, and (f) proluvium in Podbela.

(a)

Figure 3: (a) Karst sinkhole in coral limestone represents a depositional place for significant amount of weathered material; Robidišče in the background. (b) Limestone breccia in Nadiža valley south of Logje is typically bedded. (c) Diamicton near Breginj, in the background Mt. Stol built of Dachstein limestone. (d) Alluvial fan sedi-

ments form gentle slopes above Kred polje, which is composed of fine grained alluvial sediments which promote to alluvial sediments near Nadiža River channel.

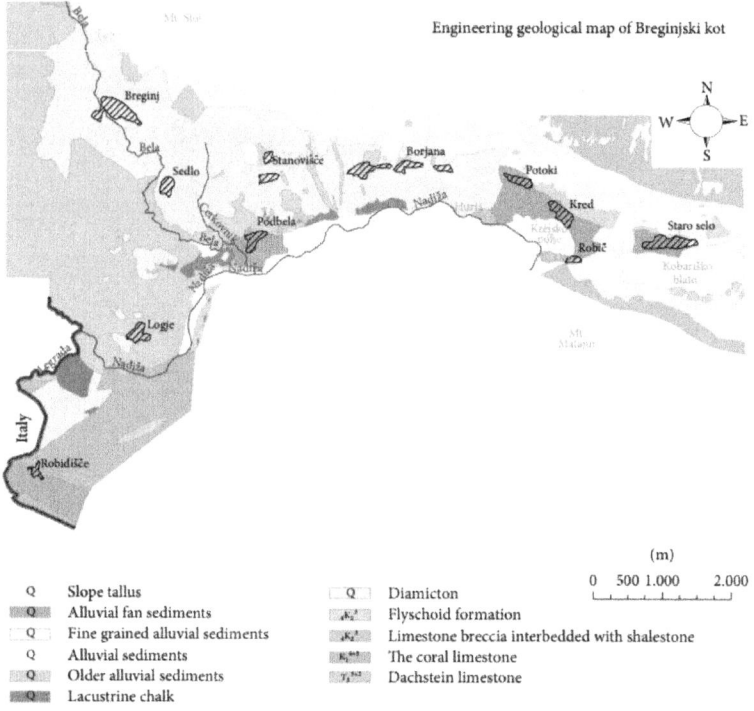

Figure 4: Engineering geological map of Breginjski kot.

Dachstein limestone (Figure 2(a)) from Late Triassic is the oldest formation in the studied area. It forms Mt. Stol, Mt. Matajur, and Der Hill west of Robič. Rock has a light grey colour and a crystalline texture. Dachstein limestone is characterized by Lofer development which consists of three parts. Cyclothem of B member contains stromatolites whereas cyclothem of C member contains well-preserved Megalodontidae shells. Cyclothem of A member consists of basal breccias but was not observed in the field. Weathered material thickness was estimated to be less than 0.5 m.

The Coral limestone of Early Cretaceous is situated in a small part of the examined area near Robidišče. Coral limestone is fine grained, massive and grey in colour. Outcrops are rare because they are overlain by significant thickness of weathered material. Hence it was distinguished from limestone breccias by numerous karst sinkholes which determine its morphology (Figure 3(a)). Moreover the weathered material thickness can be altered promptly as a consequence of a karst relief. On the slopes less than 1 m of weathered material is expected, whereas more than 5 m of weathered material is estimated for the flatland.

Limestone breccia interbedded with shalestone (Figure 2(b)) from Upper Cretaceous represents mainly the area from Napoleon Bridge west of Podbela to Robidišče (Figure 3(b)). Limestone breccias are of light grey or grey colour. Breccia is clast supported and composed of angular to subangular clasts which range from 2 mm to 1 m in size. Breccia formation is mostly bedded, rarely massive. If bedded, its thickness ranges from 10 cm to several meters in size, moreover brownish green shalestones are often interbedded. Formation generally dips to the north. On the slopes less than 1 m of weathered material is expected, whereas from 2 m to 3 m of weathered material is estimated for the flatland.

Flyschoid formation (Figure 2(c)) is the next formation of Upper Cretaceous and is situated mainly west of Breginj-Sedlo-Podbela line, while plenty outcrops are exposed in erosion windows of Mt. Stol and Mt. Matajur, especially in ravines with high-energy creeks. The Flyschoid formation is divided into series of rocks: shalestone, sandstone, and breccia, which are stratified and in sequence one after the other. Shalestone is dark grey and appears with thickness of beds ranging from 1 cm to 5 cm. Sandstone, which mostly consists of carbonate minerals, is grey with thickness of beds ranging from 5 cm to few ten centimetres. Breccia is grey to dark grey and consists of carbonate clasts ranging from 0.5 cm up to few meters. Coarse grained breccia is clast supported. Thickness of breccia beds ranges from 1 m to few ten meters. Flyschoid formation dips generally towards the north, moreover thin bedded series in particular are often found folded. Thickness of weathered material varies from 0.5 m to 1 m on the slopes and from 2 m to 3 m on the flatland.

Diverse Quaternary sediments in Breginjski kot are the result of postglacial era when water erosion and other transformational processes formed specific relief. Sediment deposits are mostly abundant on the slopes of Mt. Stol where most settlements are located.

Diamicton (Figure 2(d)) is an unlithified sediment which origins from glacial till mixed with other Quaternary sediments and transported by mass flow. It is deposited on the slope of Mt. Stol (Figure 3(c)), confined by Potoki to the east, a large area of flyschoid formation to the west and Nadiža River to the south. In addition some erosion patches can be found between Sedlo and Logje. According to the estimation on the field, diamicton is composed of more or less subrounded grains which are very poorly sorted. 70% of grains represent fine gravel, sand, and silt fractions, which ratio can vary significantly on studied locations. The remaining 30% is mainly gravel which ranges from

1 cm to 5 cm in size and boulders which ranges from 20 cm to several meters in size. Fragments consist of carbonate in general. Besides, there are many clastic fragments and diamictite ones, which are rather scarce. Glacial striations on fragments were not observed. Diamicton thickness varies significantly, for instance: along the Nadiža River only a few meters can be reached, in vicinity of flyschoid formation erosion windows it can vary between 5 m and 10 m, whereas away from ravines it can reach more than 20 m. The thickness depends on paleorelief and selective sedimentation processes, therefore a large amount of material has been deposited there. For instance in Cerkovnik creek flyschoid formation is not exposed, despite more than 20 m deep eroded ravine. According to ravine morphology, the biggest thickness is expected in the vicinity of Breginj.

Diamictit has the same origin as diamicton but it is lithified. It is not presented on the engineering-geological map, because it is scattered around the Diamicton area in small patches of about 10 m^2 in size. It consists of around 70% of gravel and boulder sized clasts which are matrix supported. The latter is composed of silt, sand and fine gravel sized grains. Subrounded clasts are poorly sorted, those in the range from 0.5 cm to 5 cm prevail. Glacial striations on clasts are rarely preserved.

Lacustrine chalk (Figure 2(e)) is deposited in patches along broader zone of Nadiža River and in the largest portion besides Legrada creek. This sediment is characterised by laminated silt, grey in colour with a blue tinge which has moisture and can be slightly molded. According to Casagrande's Soil Classification System (ASTM) [21] it was defined as ML, inorganic silt and very fine sands with low plasticity. Besides laminated silt there can be less than 5% of rounded pebbles found, mainly ranging from 1 cm to 5 cm and rarely to 10 cm. Lacustrine chalk is covered by diamicton; main evidence can be found near to Hurja farm.

Older alluvial sediments can generally be found between Kred and Podbela up to 320 m above sea level as erosion patches around 40 m^2 in size and are consequently not thicker than 5 m. They include conglomerate, but sands and gravel are more rare. First is clast supported conglomerate which is good sorted. The clasts, which range from 0.2 to 5 cm in size, are well rounded and consist mainly of carbonate and flyschoid fragments.

Alluvial sediments are located along former and actual steam channels of Nadiža River and Bela creek and in alluvial terraces of Kred polje (Figure 3(d)) and Podbela area. The volume ratio of sand and pebbles is 60 : 40. Fragments, which consist of carbonate and flyschoid formations, are well-rounded and moderately sorted. Thickness of alluvium deposits is estimated to be more than 10 m.

Fine grained alluvial sediments are deposited in Kred polje (Figure 3(d)) and Kobariško blato which represent a sedimentary basin between Mt. Stol and Mt. Matajur. The flat relief induces rare outcrops, whereas some exposures were seen in old melioration trenches. Some 4 m pits, which were excavated in the frame of geological investigations for planned wastewater treatment plant, were also examined. A particular sediment found in trenches and pits is characterised by silt of bluish grey colour. According to ASTM it is defined as ML, inorganic silt with low plasticity. On the foothills of Mt. Stol the formation is overlain by alluvial fan sediments.

Alluvial fan sediments (proluvium) (Figure 2(f)) are deposited by torrential creeks on the foothills of Mt. Stol from west of Staro Selo to Potoki (Figure 3(d)). In Podbela alluvial fan sediments are found as well. The formation consists of flyschoid and carbonate fragments, whereas the ratio depends on geological setting of hinterlands. Proluvium is poorly sorted since it consists of fragments generally ranging up to 10 cm, of which up to 40% is fine sand and silt. Boulders ranging up to 60 cm can also be observed. In addition, fragments are medium rounded. Thickness of formation is estimated on less than 20 m.

Slope talus (gravel) is deposited below steep slopes of Mt. Stol and Mt. Matajur. The lower boundary of the slope talus on the slope of Mt. Stol is increasing in elevation from east to west, whereas on the slope of Mt. Matajur it is decreasing in the same direction. Gravel consists of limestone and dolomite fragments of Late Triassic. They were distinguished by the reaction with dilute HCl. Besides gravel sized fragments, also significant amount of cobbles and some boulders can be found. Additionally patches of lithified slope tallus (gravel breccia), which have the same origin as unlithified, can be observed, but they are not presented in the map due to their negligible size and dispersion.

The described lithologies are presented on the detailed engineering geological map (Figure 4), which is the basis for the seismic microzonation of the studied area.

INTENSITY INCREMENTS SEISMIC MICROZONATION

According to the old seismic hazard map of Slovenia for a 500-year return period [14] Breginjski kot is estimated to have IX MSK intensity. It is well-known that variations in local site characteristics have influence on observed earthquake intensity. Medvedev [3] proposed a classification that improves estimation on intensity considering engineering geological conditions. According to this classification, the soil is divided into three categories. The 2nd category corresponds to the medium stiff ground, which agrees with the given intensity on the seismological map [14], whereas for the 1st category an intensity decrease, and for the 3rd category an intensity increase is expected.

According to this scheme Breginjski kot is subdivided into three categories (Table 3) and presented in the map (Figure 5). In addition, engineering geological conditions, topography, and soil saturation are given descriptively in the text below, according to Medvedev [3] and Mayer-Rosa and Jimenez [22].

Table 3: Soil categories for seismic microzonation according to Medvedev method (after [3])

Soil category	Soil type	Intensity increment according to seismic area		
		VII	VIII	IX
I	hard rocks	VI	VII	VIII
II	medium-hard rocks, lithified Quaternary sediments, alluvial sediments	VII	VIII	IX
III	unlithified Quaternary sediments	VIII	IX	X

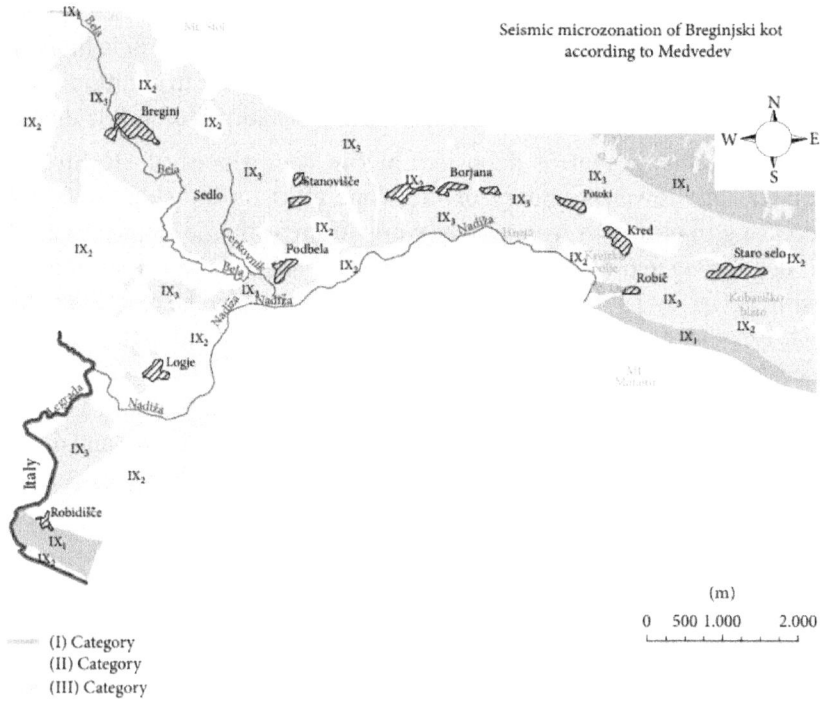

Figure 5: Intensity increments seismic microzonation of Breginjski kot according to the Medvedev method.

Category I represented by Dachstein limestone and coral limestone which are in general seismically insensitive. But, due to limestone karstification some intensity increments may be expected also for these rocks. In addition, terrain subsidence is expected because of cave collapses. Severe earthquakes may also trigger rockfalls and rock slides which can endanger settlements or road facilities.

Category II is characterised by flyschoid formations, limestone breccia interbedded with shalestone, older alluvial sediments, and alluvial sediments. Besides them, diamictit and slope tallus (gravel breccia) are also within this category, but are not presented on the map because of negligible size and dispersion. Intensity increments may be significant due to the high- degree saturation of soil along Nadiža River or Bela creek and limestone breccia karstification. Severe earthquakes may also trigger rockfalls and rock slides which can endanger settlements or road facilities. Areas with increased thickness of weathered material are prone to landside occurrences. Many caves, abysses, and sinkholes are observed in the Robidišče plateau (Figure 3(a)), where the thickness of weathered material in general exceeds 5 m and can even reach 20 m in sinkholes. A local increment of intensity can be therefore expected.

Category III is represented by lacustrine chalk, fine grained alluvial sediments, alluvial fan sediments (proluvium), diamicton, and slope tallus (gravel). Intensity increments might be larger at contacts between different sediments, especially in Podbela, Krejsko polje, and Kobariško blato which are located on the foothills of Mt. Stol. Moreover, prominent topography changes like in Sedlo, which is situated in crest vicinity, can cause an additional intensity increment.

Since significant diversities in geological conditions and topography are observed in the studied area, which might cause an increase of intensity, five main reasons for site effects in different settlements of Breginjski kot are given in Table 4. Soft sediments should be considered as a complex set of depositional events of distinctive origin, hence different sediment layers exist, divided by a certain number of sediment contacts. Generally, the larger the number of contacts is, the larger the intensity increment is. In addition, prominent topography with steep slopes, sharp relief changes and abundant karst features are also reasons for an intensity increase.

Table 4: Main reasons for significant site effects in settlements of Breginjski kot

Settlement	Prevailing lithology	No. of sediment contacts	Thickness to the basement rock	Topography	Karst features
Borjana	Diamicton	1	5–10 or more	Steep-moderate slopes	no
Kred	Proluvium	2	<10 m	Flat	no
Robič	Gravel and limestone	2	<5 m	Flat	no
Potoki	Proluvium and flyschoid f.	1	0–10 m or more	moderate slopes	no
Logje	Flyschoid f.	0*	<2 m	slight slopes-flat	no
Staro Selo	Proluvium	3	>20 m	slight slopes-flat	no
Sedlo	Diamicton	1	10–20 m or more	slight slopes, sharp relief change	no
Breginj	Diamicton	1	>20 m	slight slopes-flat	no
Podbela	Proluvium	2	>20 m	flat	no
Robidišče	Limestone and limestone breccia	0*	>5 m	flat	yes

Since 2008 seismic intensity is not used in Slovenia for building design anymore, but it is still important for seismological analysis and for civil protection. Broader applicability of the method is also in comparison with similar studies. The new seismic microzonation according to Medvedev method improves the older seismic microzonation [1] in many aspects: (a) mapping was performed on a larger scale, (b) lithologies descriptions are more precise and new interpretations are made and (c) map is digitalized. Moreover, the Bovec basin seismic microzonation [19] can also be compared to Breginjski kot due to geographic proximity and similar geological structures, though many different geotechnical and geophysical data were also acquired in the Bovec basin, as well as a study based on the microtremor method [23].

EUROCODE 8 SEISMIC MICROZONATION

The new seismic hazard map of Slovenia defines the seismic hazard with the peak ground acceleration. Furthermore, local ground conditions account for the additional influence on the seismic ground motion. In Europe ground types with different soil factors are usually determined according to the seismic standard Eurocode 8 [2] and National Annex [24].

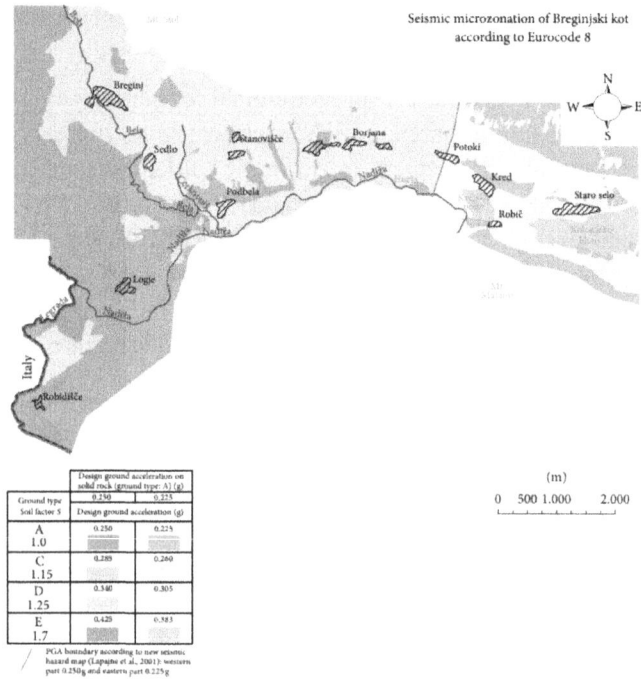

Figure 6: Eurocode 8 seismic microzonation of Breginjski kot.

According to the seismic hazard map for a 475-year return period [15], Breginjski kot is characterized in its larger western part by a design ground acceleration value for a rock site of 0.250 g and in the smaller easternmost part by 0.225 g. Ground types are classified on the basis of their geomechanical properties and weathered material thickness. Thus four different ground types, namely A, C, D, and E were assessed for the studied area (Figure 6).

Class A represents the least vulnerable ground type. It is characterised by limestone, limestone breccia interbedded with shalestone and flyschoid formation, and less than five meters of weathered material. According to the map (Figure 6) the concentration of class A is the largest in the westernmost part, though small patches are present throughout the area. Logje is the only settlement which is located entirely on class A ground. Only small parts of Robič, Potoki, and Borjana are lying on class A ground.

Class C is characterized by alluvial sediments, several tens of meters thick. The area which corresponds to the class C is situated along Nadiža River or Bela creek where no settlement is located.

Class D is characterized by diamicton, alluvial fan sediments, and slope tallus (gravel). This is the largest area in the map, which is mainly located on the slopes of Mt. Stol and Mt. Matajur. In addition it can be found in two patches between Sedlo and Robidišče. Due to its wide extent, most settlements belong to this class, for example, Staro Selo, Kred, Robič, Potoki, Borjana, Podbela, Sedlo, and Breginj.

Class E has the highest soil factor according to Eurocode 8. It comprises soil type C or D with a thickness varying from 5 m to 20 m underlain by a rock of type A. In addition, fine grained alluvial and lacustrine sediments belong to the latter type as a result of their poor geomechanical properties and heterogeneity. The area of class E includes Kobariško blato, Krejsko polje, Robidišče plateau, and some smaller patches along Nadiža valley. Robidišče is the only settlement which corresponds to the class E.

In addition to the upper classification some additional risks persist for each class, such as terrain subsidence in areas covered by limestone due to karst caves collapse and possibility for rockfalls. Final design acceleration values for each ground type are computed in accordance to the following expression:

$$\mathrm{DGA_F = DGA_A \times S \times T_{NCR}}, \tag{1}$$

where final design ground acceleration ($\mathrm{DGA_F}$) corresponds to design ground acceleration on type A ground spectral acceleration multiplied by the soil factor (S) and the factor for the reference return period ($\mathrm{T_{NCR}}$). Topographic factor is neglected in this study but it should be taken into account for important structures according to Eurocode 8 [25]. Final design ground acceleration values are shown in Table 5. According to the National Annex [24] the soil factor for ground type E is 1.7, whereas in the original Eurocode 8 standard [2] it is 1.4; for all other ground types the soil factors are the same.

Table 5: Final design ground acceleration values for each soil type in Breginjski kot.

Soil type	S	T_{NCR}	DGA_A [g]	DGA_E [g]
A	1.0	475 years	0.250	0.225
C	1.15		0.285	0.260
D	1.35	1.00	0.340	0.305
E	1.70		0.425	0.383
A	1.0	1000 years	0.315	0.280
C	1.15		0.360	0.325
D	1.35	1.25	0.420	0.380
E	1.70		0.531	0.478

DISCUSSION OF RESULTS

According to the engineering geological map of Breginjski kot, twelve different lithologies are presented and described in detail. Several of them are interpreted anew compared to previous studies (e.g., [18]).

Since typical glacial sediments are missing due to mass flows, alternative term diamicton for pseudoglacial sediments is suggested. Diamictit is also presented but only locally and in small patches. A recent study has shown that similar types of soft sediments or rocks are also identified in nearby Bovec basin [26]. Furthermore, gravel breccia was also observed in small patches similar to diamictit. Neither of them are presented in the engineering geological map but are only presented descriptively, because of their small size and sparse occurrence.

Fine grained alluvial sediments in Krejsko polje and Kobariško blato were deposited after Nadiža River had changed the current direction from the relatively wide Kobariško blato valley into the recent Nadiža valley, which is very narrow, with steep and rocky slopes [20]. Hence relatively large area was covered by swamplands and oxbow lakes where fine grained material was deposited and accumulated throughout Quaternary, although intense melioration in modern ages results in controlled river channel and thus in decreased sedimentation.

Because of detailed mapping, lithologies are defined more accurately and restricted in comparison with previous studies [20]. Therefore, their position and geometry can be better understood, thus contributing to the adequate estimation of the number of sediment contacts, their thicknesses, and abundance of karst phenomena, which may result in additional site effects

(Table 4). Moreover, thicknesses of weathered material were estimated for each rock formation, because this also has influence on seismic effects.

A significantly large thickness of weathered material is located in the Robidišče plateau due to extreme weathering of shale stone which outcrops at the edge of the plateau. The Robidišče plateau is characterized by limestone breccia and coral limestone, which includes karst features like sinkholes where the thickness of weathered material generally exceeds 5 m. Moreover, the thickness of weathered material varies significantly, since in some areas bedrock outcrops are observed. Karst phenomena like caves and abysses can also contribute to an increase of seismic ground motion. According to the geological mapping, limestone breccia interbedded with shale stone is less subjected to karst feature occurrences or such features are scarce. In the settlements of Breginjski kot small thickness of weathered material was observed especially in Logje, where outcrops of rock formation are abundant. On the other hand, the outcrops in Borjana, Potoki, and Robič were limited, thus soft sediments are more abundant. Their cumulative thickness of less than 5 m is expected. However, it should be noted that geological characteristics of sediments and their thickness are estimated from the engineering geological mapping, whereas for a more detailed analysis systematic geotechnical drilling and geophysical investigations should be performed.

The damage to buildings in settlements of Breginjski due to the 1976 Friuli earthquake was unevenly distributed, although their distance from the epicentre does not vary significantly. For comparison, the soil classification is determined for each settlement of Breginjski kot according to both seismic microzonations. A comparison between intensities of the 1976 earthquake and soil classifications is given in Table 6.

Table 6: Intensities of 1976 earthquake in different settlements of Breginjski kot and soil classification after Medvedev and Eurocode 8 from both seismic microzonation maps

Settlement	Intensity EMS-98	Soil classification	
		After Medvedev	After Eurocode 8
Borjana	VI	III, II	D, A
Kred	VI	III	D
Robič	VI	III, I	D, A
Potoki	VI+ (MSK-78)	III, II	D, A
Logje	VI-VII	II	A
Staro Selo	VII	III	D
Sedlo	VII-VIII	III	D
Breginj	VIII	III	D
Podbela	VIII	III	D
Robidišče	VIII-IX	II	E

Soil classification according to Medvedev method does not show a clear correlation of ground type to observed intensities. The main reason for this is that Medvedev soil classification does not take into account the thickness of soft sediments or weathered material. In addition it does not consider sufficiently engineering geological conditions which are annotated descriptively only. On the contrary, Eurocode 8 soil classification clearly shows a correlation between the ground type and observed intensities. This standard classifies ground types more precisely, also since it considers the thickness of soft sediments and weathered material. A better evaluation of local soil effect is therefore achieved. Based on the 1976 earthquake intensities, the settlements of Breginjski kot can be divided into two categories. The first group consists of settlements which show relatively low site effects like Borjana, Kred, Robič, Potoki, and Logje. They are located either on ground type A, or on the weathered material less than 5 m thick. The indicators of small weathered material thickness are frequent erosion windows. The second group consists of settlements which show relatively high site effects like Staro Selo, Sedlo, Breginj, Podbela, and Robidišče. They are located on soft sediments with considerable thickness. Additional causes of significant site effects in each particular settlement are described in Table 4.

CONCLUSIONS

This study represents a contribution to the evaluation of the seismic hazard in Breginjski kot, which is among the most endangered seismic areas in Slovenia. Two new seismic microzonations were prepared, because the previous seismic microzonation was based solely on the basic geological map and did not include supplementary field research, thus making it fairly inaccurate. In this study detailed engineering geological mapping of Breginjski kot in scale 1 : 5.000 was conducted. The results of mapping showed a diverse composition of sediments and rocks. Mapped units were described in detail and some of them are interpreted anew. Descriptions of mapped units also include the estimated thickness of soft sediments and weathered material. The results can be enhanced and validated by drilling and geophysical investigations, which is recommended especially in the settlements that were badly damaged during the 1976 Friuli earthquake. In addition, a microtremor method study would be appropriate to apply in the flatlands of Krejsko polje and Kobariško blato to determine soil resonance frequencies.

On the basis of engineering geological mapping, seismic microzonation was prepared according to the Medvedev method and the Eurocode 8 standard. According to the first classification the soil in Breginjski kot was divided into three categories, and according to the latter classified into categories A, C, D, and E. Two maps of seismic microzonation, based on both soil classifications, were made in the GIS platform and are therefore ready to use in urban planning. Seismic microzonation according to Eurocode 8 is more appropriate for the application, because it is in accordance with the modern seismic hazard standards and because it better correlates thickness of soft sediments or weathered material with site amplification. Moreover, ground types are classified more precisely, and thus the correlation with the geological map is better. The new seismic microzonation reveals the correlation of site effects and observed intensities. Sites located on softer ground experienced a higher degree of damage during the Friuli 1976 earthquake. Although the microzonation based on seismic intensities is not used in earthquake engineering anymore, it is still important for civil protection purposes and in seismological analysis. The new map improves the old microzonation and enables a comparison of geologically similar areas of Breginjski kot and the Bovec Basin.

ACKNOWLEDGMENT

The study was realized with the support of the research program P1-0011 financed by Slovenian Research Agency.

REFERENCES

1. R. Vidrih, M. Godec, and J. Lapajne, Earthquake Risk in Slovenia, Seismological Survey of Slovenia, Ljubljana, Slovenia, 1991.
2. CEN, Eurocode 8—Design of Structures for Earthquake Resistance, Part 1: General Rules, Seismic Actions and Rules for Buildings, European Standard, EN, 1998-1: 2004 (E), Stage 64, European Committee for Standardization, Brussels, Belgium, 2004.
3. S. V. Medvedev, Engineering Seismology, Građevinske knjige, Belgrade, Serbia, 1965.
4. B. Perniola, G. Bressan, and S. Pondrelli, "Changes in failure stress and stress transfer during the 1976–77 Friuli earthquake sequence," Geophysical Journal International, vol. 156, no. 2, pp. 297–306, 2004. G. B. Carulli and D. Slejko, "The 1976 Friuli (NE Italy) earthquake," Giornale di Geologia Applicata, vol. 1, pp. 147–156, 2005.
5. V. Ribarič, "Earthquakes in Friuli in 1976 and the short history of seismicity at the margin of Eastern Alps," Potresni Zbornik, pp. 17–80, 1980.
6. P. Zupančič, I. Cecić, A. Gosar, L. Placer, M. Poljak, and M. Živčić, "The earthquake of 12 April 1998 in the Krn Mountains (Upper Soča valley, Slovenia) and its seismotectonic characteristics," Geologija, vol. 44, pp. 169–192, 2001.

7. I. Cecić, M. Živčić, T. Jesenko, and J. Kolar, "Earthquakes in Slovenia in 2004," Potresi V Letu 2004, pp. 16–40, 2006.

8. J. Bajc, A. Aoudia, A. Saraò, and P. Suhadolc, "The 1998 Bovec-Krn mountain (Slovenia) earthquake sequence," Geophysical Research Letters, vol. 28, no. 9, pp. 1839–1842, 2001.

9. A. Gosar, "Site effects and soil-structure resonance study in the Kobarid basin (NW Slovenia) using microtremors," Natural Hazards and Earth System Sciences, vol. 10, no. 4, pp. 761–772, 2010.

10. V. Kastelic, M. Vrabec, D. Cunningham, and A. Gosar, "Neo-Alpine structural evolution and present-day tectonic activity of the eastern Southern Alps: the case of the Ravne Fault, NW Slovenia," Journal of Structural Geology, vol. 30, no. 8, pp. 963–975, 2008.

11. F. Fitzko, P. Suhadolc, A. Aoudia, and G. F. Panza, "Constraints on the location and mechanism of the 1511 Western-Slovenia earthquake from active tectonics and modeling of macroseismic data,"Tectonophysics, vol. 404, no. 1-2, pp. 77–90, 2005.

12. R. Camassi, C. H. Caracciolo, V. Castelli, and D. Slejko, "The 1511 Eastern Alps earthquakes: a critical update and comparison of existing macroseismic datasets," Journal of Seismology, vol. 15, no. 2, pp. 191–213, 2011.

13. V. Ribarič, Seismological Map of Slovenia for 500 Years Return Period, Seismological survey of Slovenia, Ljubljana, Slovenia, 1987.

14. J. Lapajne, B. Šket-Motnikar, and P. Zupančič, "Design ground acceleration map of Slovenia," Potresi V Letu 1999, pp. 40–49, 2001.

15. N. Ambraseys, P. Smit, R. Sigbjornsson, P. Suhadolc, and B. Margaris, "Internet-Site for European Strong-Motion Data," 2002,

16. B. Šket-Motnikar and T. Prosen, "Accelerations in Posočje on July 12 2004," Potresi V Letu 2004, pp. 105–113, 2006

17. S. Buser, Basic Geological Map of Yugoslavia 1: 100.000—Sheet Tolmin and Videm, Geological Survey of Slovenia, Ljubljana, Slovenia, 1986.

18. M. Ribičič, R. Vidrih, and M. Godec, "Seismogeological and geotechnical conditions of buildings in Upper Soča Territory, Slovenia," Geologija, vol. 43, no. 1, pp. 115–143, 2000.

19. S. Buser, Basic Geological Map of Yugoslavia 1: 100.000—Sheet Tolmin and Videm Explanatory Text, Geological Survey of Slovenia, Ljubljana, Slovenia, 1986.

20. A. Casagrande, "Classification and identification of soils," Transactions of the American Society of Civil Engineers, vol. 113, pp. 901–930, 1948.

21. D. Mayer-Rosa and M. J. Jimenez, Seismic Zoning: State-of-the-Art and Recommendations for Switzerland, Federal Office for Water and Geology, Swiss National Hydrogeological and Geological Survey, Berne, Switzerland, 2000.

22. A. Gosar, "Microtremor HVSR study for assessing site effects in the Bovec basin (NW Slovenia) related to 1998 Mw5.6 and 2004 Mw5.2 earthquakes," Engineering Geology, vol. 91, no. 2–4, pp. 178–193, 2007

23. SIST, Slovenian Standard SIST EN, 1998-1 Eurocode 8: Design of Structures for Earthquake Resistance—Part 1: General Rules, Seismic Actions and Rules for Buildings—National Annex, Slovenian Institute for Standardization, Ljubljana, Slovenian, 2004.

24. CEN, Eurocode 8—Design of Structures For Earthquake Resistance, Part 5: Foundations, Retaining Structures and Geotechnical Aspects, European Standard, EN, 1998-5: 2004 (E), Stage 64, European Committee for Standardization, Brussels, Belgium, 2004.

25. M. Bavec, "New temporal, and genetic determinations of some late Quaternary sediments in the Bovec basin and its surroundings (NW Slovenia)," Geologija, vol. 45, no. 2, pp. 291–298, 2002.

CITATION

CHAPTER 1
Mohammad Ali Zare Chahouki, Multivariate Analysis Techniques in Environmental Science, ISBN: 978-953-307-468-9.

CHAPTER 2
Khageshwar Singh Patel, Shobhana Ramteke, Yogita Naik, Bharat Lal Sahu, Saroj Sharma, Jutta Lintelmann, Matuschek Georg, Contamination of Environment with Polycyclic Aromatic Hydrocarbons in India, http://dx.doi.org/10.4236/jep.2015.611111.

CHAPTER 3
Roberta Pirastu, Pietro Comba, Ivano Iavarone, et al., "Environment and Health in Contaminated Sites: The Case of Taranto, Italy," Journal of Environmental and Public Health, vol. 2013, Article ID 753719, 20 pages, 2013. doi:10.1155/2013/753719.

CHAPTER 4
María Antonieta Gómez-Balandra, María del Pilar Saldaña-Fabela, Maricela Martínez-Jiménez, The Mexican Environmental Flow Standard: Scope, Application and Implementation, http://dx.doi.org/10.4236/jep.2014.51010.

CHAPTER 5

Fereshte Haghighi, Mirmasoud Kheirkhah and Bahram Saghafian, Evaluation of Soil Hydraulic Parameters in Soils and Land Use Change, ISBN: 978-953-307-468-9.

CHAPTER 6

Kevin Lo, "Energy-Related Carbon Emissions of China's Model Environmental Cities," Geography Journal, vol. 2014, Article ID 204745, 7 pages, 2014. doi:10.1155/2014/204745.

CHAPTER 7

Helen Kopnina, and Brett Cherniak, Cultivating a Value for Non-Human Interests through the Convergence of Animal Welfare, Animal Rights, and Deep Ecology in Environmental Education, doi:10.3390/educsci5040363.

CHAPTER 8

Daniel L. Childers, Mary L. Cadenasso, J. Morgan Grove, Victoria Marshall, Brian McGrath and Steward T. A. Pickett, An Ecology for Cities: A Transformational Nexus of Design and Ecology to Advance Climate Change Resilience and Urban Sustainability, doi:10.3390/su7043774.

CHAPTER 9

Giuseppe Ioppolo, Stefano Cucurachi, Roberta Salomone, Giuseppe Saija and Luigi Ciraolo, Industrial Ecology and Environmental Lean Management: Lights and Shadows, doi:10.3390/su6096362.

CHAPTER 10

Christian Mulder, Ecology and eScience, DOI: 10.1186/2192-1709-1-1.

CHAPTER 11

Yajun Song and Tianzhong Zhao, A bibliometric analysis of global forest ecology research during 2002–2011, DOI: 10.1186/2193-1801-2-204.

CHAPTER 12

Sewall BJ, Freestone AL, Hawes JE, Andriamanarina E (2013) Size-Energy Relationships in Ecological Communities. PLoS ONE 8(8): e68657. doi:10.1371/journal.pone.0068657.

CHAPTER 13

Jure Kokošin and Andrej Gosar, "Seismic Microzonation of Breginjski Kot (NW Slovenia) Based on Detailed Engineering Geological Mapping," The Scientific World Journal, vol. 2013, Article ID 626854, 12 pages, 2013. doi:10.1155/2013/626854

INDEX

A

Agriculture, forest management, and other land uses (AFOLU) 121
Animal Liberation Front (ALF) 138
Animal rights (AR) 129, 130
Animal welfare (AW) 129, 130
Animal welfare education (AWE) 130

B

Baltimore Ecosystem Study Long-Term Ecological Research Program (BES LTER) 163
Building Healthy Communities (BHC) 164
Bulk density (BD) 104, 106

C

Canonical correspondence analysis (CCA) 1
Canonical Correspondence Analysis (CCA) 1, 21
Carbonate carbon (CC) 37
Climate regulation 196, 204
Computer Aided Manufacturing (CAM) 178
Computer Integrated Manufacturing (CIM) 175
Concentrated animal feeding operations (CAFOs) 132
Cyclothem 260

D

Decision-making 205, 206
Deep ecology (DE) 129, 130, 143
Deprivation index (DI) 54

E

Earth Liberation Front (ELF) 138
Ecological network analysis (ENA) 183
Education for sustainable development (ESD) 129, 131, 143
Elemental carbon (EC) 37
Enterprise Resource Planning (ERP) 178
Environmental education (EE) 129
Environmental flow assessment (EFA) 88
Environmental Impact Assessment (EIA) 87
Environmental objective (EO) 90
Environmental protection 115, 116, 117
Eulemur coronatus 231
European Technology Platforms (ETP) 188

F

Field capacity (FC) 104
Forest ecology 205, 213, 222
Forest ecology research 205, 206, 209, 214, 215, 220, 221, 276
Foundation for Deep Ecology (FDE) 142
Frequency analysis method 206
frugivore 225, 230, 231, 232, 233, 234, 238, 241

G

Geomechanical 267
Global population 195
Global sustainability 195

I

Industrial Ecology (IE) 175, 177, 181
Information Communication Technology (ICT) 178
Integrated Water Resources Management (IWRM) 87
International Energy Agency (IEA) 115
International Fund for Animal Welfare (IFAW) 131

J

Just-in-time (JIT) 177

K

Karhunen–Loève transform (KLT) 13

L

Lean Management (LM) 175, 177, 178
Lean Production (LP) 177

M

macroecology 225, 226, 242
Madagascar 225, 229, 230, 231, 245, 248, 249, 251, 252
Mean annual flow (MAF) 90, 92
Microzonation 253, 254, 255, 257, 263, 264, 266, 269, 270, 271

Multi-response Permutation Procedure (MRPP) 8

N

National Development and Reform Commission (NDRC) 118
National Priority Contaminated Sites (NPCSs) 52
National Science Foundation (NSF) 165
Non-metric Multidimensional Scaling (NMS) 1

O

Organic carbon (OC) 37

P

Pedo transfer functions (PTFs) 103
People for the Ethical Treatment of Animals (PETA) 143
Permanent wilting point (PWP) 104
Physical input-output table (PIOT) 183
Polycyclic aromatic hydrocarbons (PAHs) 35
population energy use (PEU) 226, 229
Principal component analysis (PCA) 1, 13
Principal Coordinate Analysis (PCoA) 14
Proper orthogonal decomposition (POD) 13

R

Redundancy analysis (RDA) 1
Royal Society for the Prevention of Cruelty to Animals (RSPCA) 142

S

size class energy use (SCEU) 229
Socioeconomic position level (SEP) 55
Soil organic matter (SOM) 104
Standardized death rates (SDRs) 54
Standardized incidence ratio (SIR) 55
Standardized mortality ratios (SMR) 53

Stress ecology 195

T

Total carbon (TC) 37
Total Ordinary Volume Regime (TOVR) 92
Total quality management (TQM) 177
Total Volume for Flood Regime (TVFR) 92
Toxic equivalent factor (TEF) 43
Toyota Production System (TPS) 177

U

Urban Design Working Group (UDWG) 163